Carmen Sandhaas

Timber fasteners

a study on input parameters for the design of timber joints

BAND 39
Karlsruher Berichte zum Ingenieurholzbau

Herausgeber
Karlsruher Institut für Technologie (KIT)
Holzbau und Baukonstruktion
Prof. Dr.-Ing. Philipp Dietsch

Timber fasteners

a study on input parameters for the design
of timber joints

by
Carmen Sandhaas

Scientific
Publishing

Habilitation, Karlsruher Institut für Technologie
Holzbau und Baukonstruktion, 2024

Tag des Habilitationskolloquiums: 22. November 2023

Referenten: Prof Pierre Quenneville, PhD
Prof Erik Serrano, PhD

Impressum

KIT Scientific Publishing

Karlsruher Institut für Technologie (KIT)
KIT Scientific Publishing
Straße am Forum 2
D-76131 Karlsruhe

KIT Scientific Publishing is a registered trademark
of Karlsruhe Institute of Technology.
Reprint using the book cover is not allowed.

www.ksp.kit.edu

Print on Demand 2024 – Gedruckt auf FSC-zertifiziertem Papier

ISSN 1860-093X
ISBN 978-3-7315-1342-1
DOI 10.5445/KSP/1000167972

Preamble

Joints are a crucial part in timber structures and exhibit an innumerable diversity in terms of geometry, layout, materials, types of fastener and loading. Their mechanical behaviour is influenced by many different factors, whose impact must be properly assessed in order to develop models that predict the joints' stiffness, capacity and load-deformation behaviour. Current design models are in fact based on the translation of the observed mechanical joint behaviour into simple mechanical models, and they need empirical equations (indirect experimental data) and characteristic values (direct experimental data) as input parameters. Considering the diversity in timber joints, not all possible joint configurations including the respective input parameters can be tested. Consequently, many input parameters are based on assumptions and extrapolations, as they are established using only a limited set of experimental data, which in general are determined on single fasteners only. Without comprehensive databases, no in-depth analyses of input parameters encompassing a large variety in terms of timber products and fasteners can be carried out. Huge databases are indeed often seen as universal remedy allowing for understanding of statistical scatter and its implications on design rules, for validation of existing models, for identification of knowledge gaps and for critical reflection of past work. Such databases can furthermore point to the future, where they could be used for advanced statistical modelling and beyond, using algorithms to predict joint behaviour and potentially make simple analytical models superfluous.

In this study, comprehensive databases were assembled, focusing on assembling as various data as possible to generate as representative data as possible. All collected data stem from ***certification tests*** and hence contain parameters such as yield moment, tensile capacity, withdrawal and head pull-through capacities, as these are the only data that are available on a large scale. Consequently, the assembled databases are by no means covering all aspects around timber joints, and they are ***retrospective***. In all databases, a large and persistent scatter was observed. Bespoke series may deliver clear trends. If, however, representative data is considered, these clear trends vanish, and sources for the scatter cannot be identified. The most probable reason for this is the much larger variability of the databases in terms of materials (both fasteners and timber products) than generally used in tailor-made test series, together with some inherent information gaps of the databases, e.g. missing information on geometrical features of the fasteners that must not be recorded in the framework of certification tests. As a consequence, globally valid regression equations can be derived based on the database, but these will deliver conservative values which may lead to incorrect prediction of failure modes, and regressions that hold for bespoke test series may not be valid anymore if a whole population is considered. An obvious answer to the discussed issues with data quality is to carry out more tests. This is, however, not expected to alleviate the situation as scatter is expected to be persistent, because many influencing factors cannot be recorded easily, e.g. human influence relating to specimen preparation and test execution, the influence of fastener production as well as the influence of the timber material directly around the fastener. There will be no progress in knowledge if only the analyses are getting more complex, based on incomplete datasets, without comprehension and validation of the physical basis.

The study was developed and written during my employment at the ***Timber Structures and Building Construction*** section of ***Karlsruhe Institute of Technology***. It would not have been possible without me working here; the huge amount of assembled experimental data derives from decennia of certification work of this section, and my knowledgeable colleagues were always open for discussions. Particularly, I want to thank Hans Joachim Blaß, Matthias Frese, Rainer Görlacher, Philipp Dietsch and, above all, ***Henning Kunkel***, who is our specialist for dowel-type fasteners and initiated many of the questions discussed in the following chapters. A special thank you goes to my two external examiners, ***Pierre Quenneville*** and ***Eric Serrano***.

My last thanks go to my family and friends for everything they brought into my life.

All databases are available upon request, and any results and conclusions derived processing these databases are the responsibility of the user.

Contents

1 Introduction

1.1 Problem description

Joints are a crucial part of timber structures and exhibit an innumerable diversity in terms of geometry, layout, materials, types of fastener and loading. Their design is, accordingly, pivotal for timber engineers. Current design rules[1] for timber joints are based on mechanical models, which need empirical equations (indirect experimental data) and characteristic values (direct experimental data) as input parameters. In Europe, the so-called "European Yield Model" (*EYM*) is used to design joints with laterally loaded dowel-type fasteners[2]. These design models encompass for instance empirical models to determine embedment strength and experimentally determined values such as withdrawal parameters. Considering the diversity in timber joints, not all possible joint configurations including the respective input parameters can be tested. Consequently, many input parameters are based on assumptions and extrapolations. For instance, it is assumed that experimentally determined values for 50 mm long ringed shank nails are valid for 100 mm long ringed shank nails of the same diameter that were not tested. Therefore, all empirical models and input parameters are established using only a limited set of experimental data, which in general are determined using single fasteners only (whereas a joint always contains more than one fastener). Without comprehensive databases, no in-depth analyses of input parameters encompassing a large variety in terms of timber products and fasteners can be carried out. The quality, validity and reliability of the currently used approach of limited sets of experimental data cannot be judged without these databases.

---

1 Here, design in accordance with Eurocode 5 (EN 1995 1-1, 2010) is considered.

2 The *EYM* is explained in many textbooks, e.g. in chapter E2 of Blaß and Sandhaas (2017). The *EYM* provides designers only with values for the capacity of a joint with one laterally loaded fastener, provided minimum distances, spacing and timber thickness are respected.

Comprehensive databases are often seen as universal remedy providing different opportunities, for example the following:

- Opportunity 1:
  Better[3] design models can be derived if more comprehensive test data is available. This may lead to general rules valid e.g. for all self-tapping timber screws or to tailor-made rules with which high withdrawal capacities of certain screw types in combination with certain timber products can be utilised.
- Opportunity 2:
  Databases allow for validation of existing rules and of extrapolations; e.g. if a model developed for a certain nail type is valid also for other nail types.
- Opportunity 3:
  Databases can reveal challenges that may not have gotten the attention they deserve; for instance challenges concerning the steel properties of fasteners used in timber engineering.
- Opportunity 4:
  Database analysis allows for critical reflection of past work, for instance to judge if experimental procedures and reporting need modifications or if testing campaigns up to now were sufficient and comprehensive enough. A thorough judgement of data quality can be performed and implications on future work can be discussed.
- Opportunity 5:
  Databases can be used as input for advanced statistical modelling, where, however, the suitability of the databases for such algorithms must be checked.
- Opportunity 6:
  Databases can be implemented in software, that can incorporate properties of many different fasteners (in particular screws) in combination with many different timber products, and as such, may make simple and general design rules unnecessary.
- Opportunity 7:
  Large and valid enough databases may even make analytical design models superfluous, as they may allow for direct determination of timber joints' stiffness and capacity without the need of an intermediate model, e.g. using aspects from machine learning. Alternatively, such a database may provide for high-quality input parameters for numerical models with which the mechanical behaviour of timber joints can be directly modelled.

---

3 "Better" may mean different things, e.g. more reliable results, results covering more application fields, bespoke models with precise boundary conditions that allow using some applications to their full potential.

1.2 Purpose of this study

A prerequisite to tackle any of the seven exemplary opportunities is, therefore, to assemble comprehensive databases and to analyse and assess them. Exactly this is the key question of this study: Can the seven opportunities be addressed once comprehensive databases are available?

Furthermore, this study wants to preserve knowledge and data that was built up at ***Karlsruhe Institute of Technology*** (*KIT*) in the last decennia, and to make these data available to other researchers. In this study, test results covering single dowel-type fasteners are considered, using data established in the last 20 years, and reflecting the variety of fasteners available on the market. The databases are examined to understand statistical scatter and its implications on design rules. Influencing factors such as density or fastener diameter on fastener properties are assessed and relationships are sought that allow for valid regression equations. Moreover, the general data quality is thoroughly discussed with the aim to identify knowledge gaps and to comment and give recommendations for testing and material standards. Consequently, the assembled databases are by no means covering all aspects around timber fasteners or timber joints in general. Here, the focus lies on assembling and analysing certification data on staples, nails and self-tapping screws – as these are the only data that are available on a large[4] scale.

1.3 Content and structure of the study

The assembled databases encompass individual test results on staples, nails and self-tapping screws, representing only few of the many aspects concerning timber fasteners and joints. Therefore, an embedding of this study and its content into the larger framework of timber joints and a discussion on aspects going beyond this study is given in chapter 2. The study then continues with ***retrospectively*** assembled experimental databases. Chapters 3 to 5 cover certification test results on staples, nails and screws. These tests include material properties such as yield moments or tensile capacities, but also properties such as withdrawal and head pull-through capacities, where the combination of fastener and wood is tested. After presentation, description and analyses of the assembled databases, chapter 6 concludes and summarises consequences of this study for research, practice and standardisation, and hence tries to give an answer to the seven opportunities formulated in section 1.1.

4 "Large" for a timber engineer, not large for a computer scientist.

2 General

2.1 Framework

The mechanical behaviour of timber joints is influenced by many different factors, whose impact must be properly assessed in order to develop models that predict the joints' stiffness, capacity and load-deformation behaviour. Current design models discussed in section 2.2 are in fact based on the translation of the observed mechanical joint behaviour into simple mechanical models, see also section 2.3. As said in section 1.1, these models require input parameters such as embedment strength or withdrawal parameters, for which information on their statistical properties must be available in order to guarantee safe enough design equations. In order to advance existing joint design procedures, therefore, enough data must be available for varying joint typologies. Depending on the used materials, geometry, layout and loading directions, different design models may exist or the input parameters may vary in terms of their type or value. A prerequisite to good models is that as many influencing factors, be they quantifiable or not, and as many relevant joint typologies as possible must be known. Indeed and as already stated in the introduction, joints exhibit an innumerable diversity in terms of geometry, layout, materials, types of fasteners and loading. Different materials with different cross-sections can be joined with a varying number of fasteners that are arranged in various patterns. The fastening of the members can be realised with carpentry joints such as step or dovetail joints, with surface-type fasteners such as nailing plates or connectors or with dowel-type fasteners such as nails, bolts, screws or glued-in rods. These joints can then be loaded for example in tension or in bending, parallel to the grain of the members or under an angle to their grain, with the (dowel-type) fasteners loaded parallel or perpendicular to their axis or in a combination of both. The used materials can be different timber products made of soft- or hardwood such as solid or glued-laminated timber, laminated veneer lumber or wood-based panels, or timber members can be fastened to other members made of metals or concrete.

This study is presenting and analysing databases with information on staples, nails and self-tapping screws, where ***only data evaluated during the fasteners' certification process is considered***. This means that possible analyses concern only certain, well-defined and quantifiable input parameters of some dowel-type fasteners, namely staples, nails and self-tapping screws. These input parameters are listed in section 2.5. Other relevant dowel-type fasteners are dowels and bolts. These, however, are standardised fasteners, where both the geometry as well as the material properties are precisely defined. The thread of the bolts, for instance, is a metric thread and the mechanical properties of dowels are in accordance with Eurocode 3 part 1 (EN 1993 1-1, 2010, e.g. grade S235 or S355) and those of bolts are in accordance with Eurocode 3 part 8 (EN 1993 1-8, 2010, e.g. grade 4.6 or 8.8). This is turn means that both types of fasteners do not require certification testing.

The parameters listed in section 2.5 are therefore representing only some aspects around timber joints. In order to fully address timber joints and their influencing factors, many more databases should be assembled. For instance, databases of other input parameters such as embedment strength or friction are not analysed here nor are test results on timber joints included. There are various reasons for this. Firstly, these databases already exist, at least partially (Aurand and Blaß, 2021; Schweigler et al., 2019), where data from different sources were assembled. Secondly, at KIT, the only extensive databases available are those assembled in this study. This means that data from literature would need to be considered as well to assemble further comprehensive databases. This, however, introduces a major element of uncertainty, i.e. that tests were carried out at different institutions, using different protocols and people, and that, regularly, not all information concerning tests and used materials is given in literature. Finally, such databases suffer the same drawbacks as the assembled databases, e.g. incomplete data, and even more so, embedment, friction and joint test results suffer significantly more from incompleteness due to more complex test setups, unclear reporting of results and generally much less available tests.

Therefore, many gaps are left open in this study, but have to be discussed in this chapter in order to place the study within the general framework of timber joints, to give relevant references[5] for all aspects not contained here and to identify persisting open issues affecting timber joints.

2.2 Design models used in standardisation

Looking at historic or current design models helps identifying those factors that influence the mechanical behaviour of timber joints and are quantifiable so that they can be used as input parameters for design. Historic timber design standards, e.g. DIN 1052 from 1969 or 1988 (DIN 1052, 1969; DIN 1052, 1988), used the allowable stress design where safety factors for both resistance and load were accumulated in the given allowable stress per fastener type. To obtain the allowable load for a bolted joint for instance, the allowable stress – an empirically derived constant numerical value – was multiplied by bolt diameter and

5 Many of these aspects were and still are discussed extensively. This study does not claim to give a comprehensive overview. Only recent publications are cited or those containing a comprehensive literature review.

timber member thickness. For joints with laterally loaded nails instead, the allowable load per shear plane was depending only on the nail diameter. Hence, no mechanical models were used to predict the joint capacity. Also joints with axially loaded nails were already covered, where the given allowable withdrawal stress was multiplied with penetration length and nail diameter to obtain the allowable withdrawal load per nail. Finally, in DIN 1052:1988, an interaction criterion for nails loaded both laterally and axially was included. Analogously to today, minimum spacing, end and edge distances and penetration lengths were a prerequisite to applying these design rules that were hence based on experimentally derived "allowable stresses" as input parameters, together with some geometric data such as diameters and penetration lengths.

Joints with laterally loaded dowel-type fasteners

In many modern timber design codes, the basic design model for joints with laterally loaded dowel-type fasteners is the European Yield Model (*EYM*), which itself is based on Johansen's model (1949), and including the rope effect stemming from increased capacity due to friction in the shear plane[6]. This basic design case calculates the load-carrying capacity of a joint per fastener and per shear plane, and uses simple mechanical models thoroughly described in textbooks (Blaß and Sandhaas, 2017). This development of mechanical models based on observed test results was a major step forward in comparison to older codes[7]. Necessary input parameters for the *EYM* are embedment strength f_h, yield moment M_y, withdrawal capacity F_w, head-pull through capacity F_{head}, tensile capacity F_t and friction coefficient μ of the shear planes[8]. As in the historic codes, prerequisite to the applicability of the *EYM* are large enough end and edge distances, spacing and timber thickness, so that the assumed failure modes with rigid-plastic material behaviour can develop. Examples for modern design codes using the *EYM* are the current Eurocode 5 (EN 1995 1-1, 2010) and the draft of the New Zealand Standard NZS AS 1720.1 (2018)[9].

It became swiftly clear that the prescribed minimum spacing and distances are not sufficient to prevent brittle failure modes in joints with more than one fastener and that, hence, the *EYM* must be amended to account for group effects. A first step was to introduce the effective number of fasteners n_{ef} that reduces the actual number of fasteners in a row in joints

6 Other effects, e.g. resistance to withdrawal along fastener axis and surrounding timber, leading to increased capacity are (currently) not considered, as these additional rope effects need large displacements to be activated.

7 It is, however, depending on the point of view if the introduction of more complex models in design is a step forward in comparison to the older simple rules.

8 For combinations of high density hardwoods with lower strength dowel-type fasteners, also the fasteners' shear capacity may be needed in future (Blaß et al., 2017).

9 The basis to this draft is the Australian standard AS 1760.1., where, consequently, the *EYM* is also used.

loaded parallel to the grain (Jorissen, 1998). Later, Eurocode 5 was extended to take also block and plug shear failures into account (Annex A of EN 1995 1-1, 2010).

Still, the *EYM* with all its amendments is not able to fully cover group effects or complex loading situations. The most modern code draft NZS AS 1720.1 contains elaborate models that account for brittle failure modes. These models are mainly based on research by Zarnani carried out on joints with rivets (Zarnani, 2013)[10]. Also recent publications are presenting design models (Jockwer and Dietsch, 2018; Yurrita and Cabrero, 2019, 2021) that are aimed at predicting timber failures occurring before the joint reaches significant deformations due to embedment and plastic bending of the fasteners, and where hence the joint capacity was not reached. Other publications address reinforcement of joints against brittle failure modes[11] (Dietsch and Brandner, 2015) in order to guarantee that joint capacities can be reached. Most modern design approaches currently under development still need all input parameters also needed for the *EYM*, extended with material properties that account for the observed timber failures, i.e. tensile strength parallel or perpendicular to the grain, longitudinal shear strength or moduli of rigidity. Some models even need fracture energies, e.g. the model covering the full splitting strength of joints with small dowel-type fasteners loaded perpendicular to the grain contained in NZS AS 1720.1.

Apart from models based on analytical approaches complemented with empirical data that are part of current standards, further developments use finite element modelling to capture the load-deformation behaviour and load distribution of joints and are intended to be incorporated in future standards. The most practice-oriented class of FE models covers so-called *BOF*-models (nonlinear beam on nonlinear foundation models), that are able to trace the full load-deformation path of joints, and hence also the stiffness (Schweigler et al., 2021)[12]. Therefore, *BOF*-models need load-deformation information for embedment and fastener bending together with information on friction or resistance in the shear planes and along the fastener axis.

Joints with dowel-type fasteners loaded in tension in their axial direction

Existing design models for joints with axially loaded fasteners (only tensile load) are simple and straightforward. Again, minimum spacing and distances must be maintained in order to guarantee the assumed failure modes. These failure modes are withdrawal of the fasteners in the point-side member, head-pull through in the head-side member or a tensile failure of

10 The design models are also part of the Canadian Standard CSA-086 (2014).

11 Reinforcement of timber joints can not only be used to mitigate brittle tensile failure perpendicular to the grain, but also to increase stiffness and capacity of joints (Bejtka and Blaß, 2005).

12 Here, a recent publication is cited. However, the use of beam-on-foundation models in timber engineering research dates back to the 50s with relevant work from Kuenzi (1955), Wilkinson (1971) and Foschi (1974). In that sense, the advent of *BOF*-models is more a rediscovery.

the fasteners, and the capacity is the minimum of the three. Also here, consequently, required input parameters are the same as for joints with laterally loaded fasteners, i.e. F_w, F_{head} and F_t. Timber design codes such as Eurocode 5 and the NZS AS 1720.1 correspond to this.

Similar to the *EYM* for joints with laterally loaded fasteners, also this basic design model cannot capture brittle failure modes occurring for groups of fasteners, and effective numbers are introduced. A particular issue here is whether the individual fasteners of a group can be loaded equally or, depending on the stiffness of the load introduction plate, if the axial load is transferred unequally. The most recent publications deal with screws and investigate group effects in joints with axially loaded screws where the screws are inserted parallel or perpendicular to the grain (Mahlknecht et al., 2021).

Shortcomings of basic design models

The basic design models presented shortly here are always related to capacities and, during their calibration, consider ultimate experimental loads[13]. Concerning stiffness, equations are available, but these are very general and cannot capture observed scatter. Above all, the basic design models (e.g. *EYM*) do not cover "real" joints with more than one fastener, but provide only one simple empirical equation per fastener type. In any case, all models must respect certain boundary conditions such as minimum distances in order to avoid brittle failure modes reached at very small deformations. This is currently the only point where deformations of joints come into play, together with further indirect rules such as reinforcing of joints in order to increase the joints' deformation capabilities. However, for robustness reasons and especially in seismic design, usually a "sufficient ductility" is required. Another example for such a requirement can be found in the EAD (2019) for screws that asks for 10 mm joint displacement before failure, if reduced spacing rules for joints with laterally loaded screws is sought (see also footnote 103). Whereas the EAD at least specifies a concrete number, i.e. 10 mm, to assess the joints' ductility, such values are usually not available. Or, in better words, there is no conventional agreement on such a value. This is extensively discussed in Ottenhaus et al. (2021), who proposed the definition of so-called performance targets in line with current performance-based design methods in earthquake engineering (Filiatrault and Folz, 2002). Ductility of joints, or rather realistic prediction of a joint's load-deformation behaviour is key to proper design in particular in case of complex and high-performance joints currently used in large engineered timber constructions. In addition, the trend of multi-storey timber buildings require correct predictions of, above all, joint stiffness. Above-mentioned novel design methods, i.e. *BOF*-models, tackle these issues (Schweigler et al., 2021).

[13] Experimental data in turn suffers from testing, measuring and reporting uncertainties, see also section 2.1.

2.3 Note on joint behaviour

In order to initiate database analyses dealing with parameters such as tensile capacity F_t, yield moment M_y, withdrawal parameter f_w or head pull-through parameter f_{head}, this section discusses some exemplary aspects of the mechanical behaviour of joints. The focus lies on single-shear timber-to-timber joints with one screw as the most basic case of a joint. A thorough understanding of the load-deformation behaviour of this basic case is key to an appropriate design, although it is reduced in its complexity (see also section 2.2). Typical load-deformation curves of the single-shear timber-timber joints with one partially threaded screw shown in Figure 2-1 on the right can be seen in Figure 2-1 on the left. Three different wood species resp. timber products were used; beech LVL, birch and spruce glulam, and all were predrilled. All results are taken from Kuck and Sandhaas (2022).

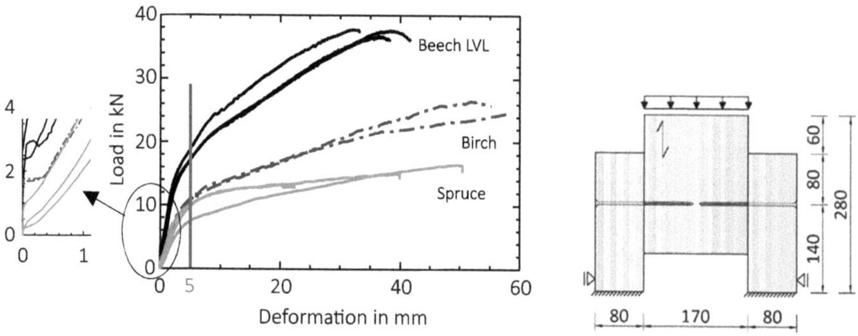

Figure 2-1 Left: Load-deformation curves of joint on the right with load per shear plane and deformation in shear plane (mean value). Right: Symmetric test specimens of timber-to-timber joints with one shear plane (all in mm). (Kuck and Sandhaas, 2022)

The tests shown in Figure 2-1 were particular as not all tests were loaded until failure, but stopped as soon as certain predefined deformation levels, 5 mm, 10 mm and 25 mm, in the shear plane were reached. The levels were chosen such that three distinct sections of the general load-deformation curve could be investigated: 5 mm represents the transition between elastic and plastic behaviour[14], 25 mm is defined as deformation at failure, and between 5 mm and 10 mm, the rope effect is assumed to be fully developed. Afterwards, the specimens were opened to investigate the deformation status. Opened specimens are shown in Figure 2-2, where the group on top shows the specimens from Figure 2-1, and the group at the bottom shows analogous tests with screws with a washer head (not predrilled).

[14] Looking at Figure 2-1 on the left, the transition between elastic and plastic behaviour is at smaller deformation levels. However, at deformations smaller than 5 mm, no (plastic) deformations could be observed (with the naked eye) when opening the specimens; see also Figure 2-2.

Figure 2-2 shows only plastic deformations as any elastic parts will have been released during the opening process.

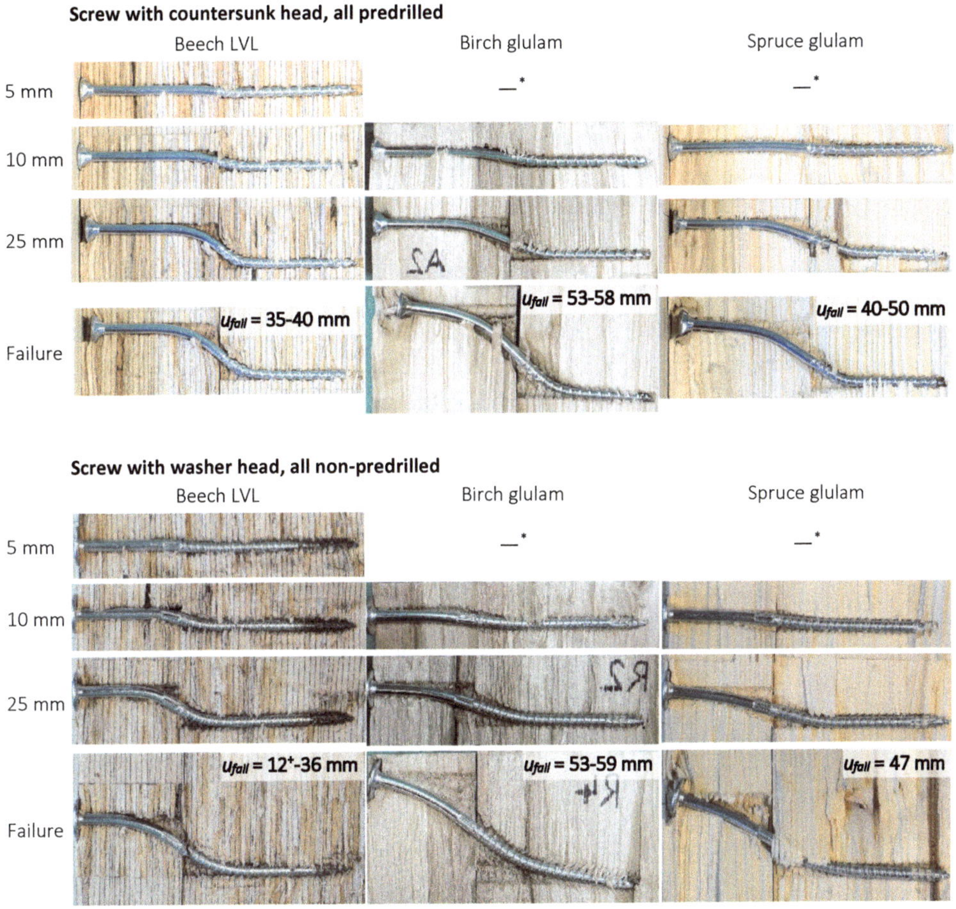

* Tests with birch and spruce glulam were not stopped at 5 mm. + Early screw failure (rupture in shear plane)

Figure 2-2 Opened specimens after the tests. The rows show the deformation status at the given deformation levels. Top: Joints with 8 mm screws with countersunk heads (corresponding to Figure 2-1). Bottom: Analogous tests with 8 mm screws with washer heads (with different cross-sections and without predrilling). Mean densities were $\rho_{spruce} = 463$ kg/m^3, $\rho_{birch} = 629$ kg/m^3 and $\rho_{beechLVL} = 814$ kg/m^3. (Kuck and Sandhaas, 2022)

Various statements can be made considering Figure 2-1 and Figure 2-2:

- The first deformation level of 5 mm did not show any perceptible plastic deformations in the birch and spruce joints and only slight deformations in the beech LVL joints, although Figure 2-1 on the left clearly reached the nonlinear range at 5 mm deformation. It is realistic to assume that micro-damage (i.e. wood crushing underneath the screws and/or micro-splits perpendicular to the grain) occurred at low deformation levels, which was not visible to the naked eye.
- Of the three species, beech LVL joints reached the highest load-carrying capacity F_{max} albeit at smaller deformations. Load levels should hence always be given in conjunction with deformation levels.
- The initial part of the load-deformation curves displayed a vertical line, which is most probably caused by static friction induced by tightening of the timber members through the partially threaded screws. Kuck and Sandhaas (2022) looked further into this by modifying the insertion process, e.g. by loosening the screws with one half-turn backwards after the insertion process.
- Concerning the stiffness, it is hard to determine a value for K_{ser} due to the initial vertical load increase (previous bullet point) and the subsequent nonlinear nature of the load-deformation curves. As long as no clear conventional method exists how to determine K_{ser} for such load-deformation curves, any derived values will be subjective[15].
- Clear embedment deformations can be seen in all tests, where the embedment length is increasing with increasing deformations. Embedment strength values and thus timber density are hence indeed one of the crucial input parameters for joint design, where, however, deformation levels of the joints certainly influence the value for the embedment strength f_h.
- The plastic hinge in the smooth shank showed larger bending radii than the hinge in the thread. This is due to the larger diameter of the smooth shank with consequently higher yield moment.
- The screw with the countersunk head was pulled in the timber, whereas the screw with the washer head displayed a higher resistance against head pull-in. This led also to a straightening of the plastic hinge (larger bending radius) on the head side of the screw (smooth shank part). The head pull-through capacity F_{head} is thus an important parameter for design, where however admissible deformation levels should be defined, in particular for screws with countersunk heads.

[15] The regulations in accordance with EN 26891 (1991) are not sufficient to guarantee a uniform determination of the stiffness. Above all, the 10%-40% rule often leads to confusion, in particular for joints with an early onset of ductility.

- Depending on the density, the bending radii of the screws differ, with larger bending radii at smaller densities. This has implications on the yield moment M_y that is needed in design (see also section 5.3.2).
- All joints showed a strong rope effect, i.e. a steep increase in the nonlinear phase. Together with the head pull-through resistance, the withdrawal capacity F_w plays a major role. To evaluate the point at which F_w starts to evolve, information on the withdrawal stiffness would be needed.
- It is notable that a steel failure occurred in the beech LVL shown in Figure 2-2 at the bottom. This failure occurred at large deformation levels and can be explained with the MNV-interaction presented in Blaß et al. (2017).
- Looking e.g. at the deformations at failure, the complexity of the rope effect gets clear, but also its dependency on the deformation level of the joint. Contributions to the load-carrying capacity caused by friction in the shear plane – this is the rope effect accounted for in current design (EN 1995 1-1, 2010)[16] – will develop at an early stage. This is also confirmed by the third bullet point, as already the tightening of (partially threaded) screws will press the timber pieces together so that friction in the shear plane will evolve, going from static to dynamic friction. For this, however, an anchoring of the screw inside the timber is needed, which is provided by the head pull-through and the withdrawal resistance. With this, also a rope effect due to normal forces along the screw axis will develop. Finally, also the distortion resistance of the screw will lead to a contribution to the rope effect, as the screw axis tends to be straightened at higher deformations, leading to larger bending radii (see also Schweigler et al., 2021).

To conclude it can be said that all well-known input parameters for joint design in accordance with e.g. the *EYM* could be identified by simply looking at Figure 2-2. Figure 2-2 however also highlighted how complex the problem to determine these input parameters is, especially for those properties that depend on a deformation level – which all of them do[17]. For the head pull-through resistance for instance, the dependency of the F_{head}-value on the reached deformation is obvious, especially for screws with washer heads inserted at 90° to the grain that activate compression perpendicular to the grain. But also the value needed for M_y is difficult to assess as e.g. the fastener geometry is not uniform along the fastener axis and because different bending radii develop (depending on steel grade, timber density, diameter), which will influence any experimentally evaluated value for M_y (see also section 5.3.2).

16 The rope effect in the current Eurocode 5 was derived for the undeformed state, Bejtka and Blaß (2002). Svensson and Munch-Andersen (2018) derived equations based on the deformed state.

17 Also the withdrawal depends on the deformation state, please refer to sections 2.4.2 and 2.4.3 for this.

2.4 Influencing factors

2.4.1 General

Having looked at current design models and exemplary test results, the most obvious influencing factors[18] are known; namely the ***quantifiable input parameters*** yield moment M_y, withdrawal capacity F_w, head-pull through capacity F_{head}, tensile capacity F_t , embedment strength f_h and friction coefficient μ of the shear planes[19]. There are many more influencing factors, however. Still one of the most thorough discussions on factors influencing the load-deformation behaviour of joints with dowel-type fasteners was presented in Werner (1993), where the focus lay on embedment strength, bending and axial resistance of the fastener and splitting tendency of the used species. Moreover, also factors influencing the embedment strength such as density, fastener diameter, loading direction, surface roughness of fastener, moisture content, predrill diameter and reinforcement were discussed (Werner, 1993). In general, depending above all on the joint's geometry, the loading and the number and layout of fasteners, the mechanical behaviour of joints in terms of load-deformation can be very different ranging from "highly ductile to very brittle". Indeed, Werner's studies encompassed mainly joints with one fastener, and joints with more than one fastener were only covered by numerical studies investigating the influence of only one factor, i.e. varying spacing parallel to the grain.

A joint with one fastener may behave ductile developing embedment deformations and plastic hinges, but it may behave brittle if more than one fastener is used, even if timber thickness and minimum spacing and distances are maintained[20]. Consequently, clear boundary conditions must be set when discussing influencing factors, as more factors must be discussed when considering joints with more than one fastener. These encompass mechanical properties of the used timber products such as moduli of elasticity, tensile strength perpendicular to the grain, longitudinal or rolling shear strength[21]. Considering mechanical properties as influencing factors complicates life considerably, as they show a large scatter and are more difficult to classify as they depend mainly on wood characteristics such as knots. At the same time, only the properties of the wood surrounding the fasteners is of importance for joints. As the probability will be lower that there are knots in the joint area volume in comparison to the whole timber member volume, knots will be less important than is reflected in the strength class system. In other words, the local and not the global

---

18 One of the most important influencing factors in general is the density.

19 F_t, M_y, F_w and F_{head} are part of the databases of this study, f_h and μ are discussed in sections 2.4.2 and 2.4.3.

20 This is why e.g. the effective number of fasteners was introduced in design, see section 2.2.

21 This is reflected in current design rules for block shear or in recently developed models that cover brittle failure modes, see section 2.2.

wood properties are expected to be governing, which makes the use of experimentally determined material properties, determined on large volumes, for design or performance prediction of joints complicated (which, however, are still subject to a volume effect, analogously to timber members). For instance, the fine differentiation of densities in the current strength classes is not relevant for the determination of the embedment strength (Sandhaas, 2012). Furthermore, properties such as shear strength or stress interactions are very difficult to assess experimentally (Sandhaas, 2012). Another good example for this is the material property "fracture energy", that is used e.g. in NZS AS 1720.1 (see section 2.2). Fracture energy is difficult to determine experimentally, subject to very considerable scatter, and generally, a constant fracture energy is used, where however a valid hypothesis would be a non-constant value for heterogeneous materials (Sandhaas, 2012).

Other, additional influencing factors exist that are not (yet) quantifiable, where some are obvious and others may not yet be known. Obvious factors include manufacturing issues such as geometric tolerances or borehole quality[22]. Indeed, load distribution within a group of fasteners is of crucial importance, which makes investigations concerning e.g. borehole precision or stiffness properties much more important for joints with more than one fastener. Another obvious influencing factor is the presence of fastener coating, whose effects are already utilised for staples that are resin-coated to increase withdrawal resistance, or for screws that are coated to decrease insertion resistance. Obviously, a clear distinction between influencing factors concerning only joints with one or those with more than one fastener is not always possible. For instance, a possible minimum end distance of a fastener may be depending on the used wood species and its mechanical properties. The distance may be larger for species prone to splitting[23] such as fir (*Abies alba*) and less large for "ductile" species such as beech (*Fagus sylvatica*), above all if the fastener is inserted in tangential direction or at 45° to the annual rings of beech (Sandhaas, 2012).

2.4.2 Embedment

Embedment strength f_h and yield moment M_y are the two non-geometric input parameters of the Johansen model, which forms the basis of the *EYM*. In the model, both parameters are assumed to be rigid-plastic, and the embedment stress is assumed to be evenly distributed. Embedment strength is determined experimentally, where two main test setups are used, the full hole and the half hole test (Franke and Magnière, 2014). In both test setups, the fastener shall remain straight, which is why the timber thickness for the full-hole test setup is reduced in order to avoid any fastener bending. Embedment tests show a nonlinear

22 See e.g. Dias (2005) Figure 4-31, where different qualities of drilled holes in spruce, maritime pine and chestnut are shown.

23 The minimum timber thickness in accordance with Eurocode 5 is larger than for species not prone to splitting such as pine (*Pinus sylvestris*).

behaviour[24], and hence, the embedment strength value is a conventional value that depends on the definition of the deformation at which the strength shall be evaluated. This is also discussed in Franke and Magnière (2014), who could show that different test setups and evaluation methods lead to different embedment strength values. An additional issue that comes into play is the accuracy and location of displacement measuring, as is the case in any other test setup where load-displacement information is derived. Furthermore, embedment tests do not reflect the real boundary conditions in joints, where in the case of failure modes with one or two plastic hinges per shear plane, the fastener is rotating in the timber leading to uneven stress distribution and complicated stress states. Additionally, during testing, the timber underneath the fastener can deform freely on the lateral specimen sides, which is not the case in joints, where such a free lateral deformation in the shear plane – where the high embedment stresses are – is not possible. Other issues during testing are boundary conditions in particular for different load-to-grain angles as was investigated by Schweigler et al. (2017). They showed that constrained loading led to higher embedment stresses compared to unconstrained loading. Also Lemaître (2020) studied these effects and measured trajectories of fasteners in timber that are different for different load-to-grain angles. Another question is to which extent simultaneous loading in embedment and withdrawal influences the individual parameters f_h and f_w, where first studies were carried out by Blaß et al. (2006), and which will occur in real joints, certainly once joint deformation has occurred. Indeed, already in the 80s, biaxial tests on nailed joints were carried out to investigate how joints under simultaneous combined loads behave (Ehlbeck and Siebert, 1984). All this highlights the difficulties with transferring embedment test results to real joints that are usually much larger both in terms of number of fasteners and cross-sections of members to be joined.

Notwithstanding these testing and interpretation issues, embedment tests are crucial for proper design of timber joints, and the embedment strength must be assessed for any new fastener type, wood product or species. Moreover, the aim of test programmes is exactly to find answers to the above questions. Consequently, innumerable publications from all over the worlds exist that document embedment test results. Recent examples for these cover different wood products such as Japanese plywood (Ogawa et al., 2019) or different fastener types such as self-tapping screws with a partial thread (Khan et al., 2021). As in every experimental work, documented test results stemming from different sources cannot easily be compared. Test procedures may differ and not always, all relevant information is given such as where displacements were measured or at which displacement the embedment strength was determined. Sometimes, machine displacement is considered during embedment tests, which introduces additional measuring errors. Or, different scopes were formulated that led to slightly different test setups. The work from Schweigler et al. (2017) discussing the effects of different boundary conditions on test results can be cited again to illustrate this. All this,

[24] This statement does not hold in all cases. In particular high-density products may split at an early stage, and often, tests are carried out using a reinforcement to reach larger deformations, see e.g. Sandhaas et al. (2013).

in turn, makes the assembly of large databases stemming from different testing bodies unreliable. This is the reason why no database with embedment tests was assembled here, as no significant amount of tests was carried out at KIT. The generation of databases containing embedment test results in terms of load-displacement curves, however, is of utmost importance particularly with the advent of the *BOF*-models. In fact, Schweigler et al. (2019) assembled a database[25], where a clear focus was on the comparability of results, in order to generate parameterised load-displacement curves needed for *BOF*-modelling. The most recent comprehensive literature review complemented with full-hole and half-hole tests was published by Ottenhaus et al. (2022).

2.4.3 Friction

Friction is an important contributor to the load-carrying capacity of joints, but also one of the most difficult to quantify and to guarantee during the lifetime of a structure. In current design, friction is considered when calculating the capacity of joints with laterally loaded fasteners that are able to transmit normal forces along their axis and where a failure mode with one or two plastic hinges per shear plane occurs. In the design equations e.g. of Eurocode 5 (DIN EN 1995 1-1, 2010), the static[26] coefficient of friction corresponds to the factor ¼ = 0.25 with which the capacity $F_{ax,Rk}$ in direction of the fastener axis is considered in the *EYM*[27]. The addition of $0.25 \cdot F_{ax,Rk}$ to the Johansen part of the design equations – the so-called rope effect – considers an increase of the load-carrying capacity due to friction in the shear plane; i.e. between the timber surfaces in a timber-to-timber joint or between timber and steel plate in a timber-to-steel joint. Obviously, the choice of a coefficient of friction is linearly influencing the calculated capacity; if 0.5 had been chosen instead of 0.25, the additional capacity would be two times as high. A proper choice for the coefficient of friction is thus crucial in order to obtain reliable design values.

The importance of correct coefficients of friction is long known, but until recently, usually values from literature were considered and only few friction tests were carried out. The most comprehensive literature reviews are contained in two recent publications (Aurand and Blaß, 2021; Dorn et al., 2021), where Aurand and Blaß assembled a database containing about 1000 test results each of timber on timber and timber on metal. Both authors carried out own friction tests, for two different scopes. The aim of Aurand and Blaß (2021) was to obtain optimised joints with higher stiffness and higher capacity due to increased friction in the shear plane by surface modification. Friction tests helped to identify suitable surface modifications. Dorn et al. (2021) were interested in generating correct coefficients of friction for use in numerical simulations. In both publications, factors influencing the coefficient of

---

[25] This database includes the tests assembled in Sandhaas et al. (2013).

[26] Here, only the static coefficient of friction is discussed.

[27] For more background, please refer to textbooks, e.g. Chapter E2.6 of Blass and Sandhaas (2017).

friction are extensively discussed, with surface roughness being most important. Test setups differed as shown in Aurand and Blaß. In addition, moisture content and contact pressure influence coefficients of friction, albeit to a lower extent. Also grain orientation, sliding speed and density were varied, with no clear conclusions. In general, observed coefficients of variation are large. Long-term effects such as changes in contact pressure or to the surface conditions e.g. due to corrosion were not studied. In other words, it is not known if the initially chosen coefficient of friction remains the same during the service life of the structure. Moreover, friction in the shear planes of joints is assumed to be constant although it is a safe assumption that friction coefficients change depending on the joints' state of deformation.

Up to now, only friction in the shear plane due to the rope effect was discussed. However, friction exists also in the contact surfaces between fastener and timber. For fasteners with threaded shanks, e.g. self-tapping screws, or fasteners with washers, e.g. bolts, this is trivial as these fasteners possess a strong withdrawal resistance that could be considered being a form of friction. Key here is that a joint with laterally loaded fasteners needs a certain deformation so that the fasteners are inclined and can develop normal forces along their axis. Schweigler at el. (2021) described these different forms of the rope effect, and Svensson and Munch-Andersen (2018) derived the Johansen equations for the deformed state so that this additional rope effect due to friction between fastener and timber along the fastener axis can be considered as well. However, independently of the shape of the shank, friction will be present as soon as a fastener is inserted in timber and shows a certain resistance against withdrawal. Indeed, Blaß et al. (2017) postulated that also joints with smooth dowels show a significant rope effect if they can develop enough deformation.

The question now is how a coefficient of friction for these cases can be determined. Rodd (1973) and Sjödin et al. (2008) could show that the surface roughness of fasteners has a considerable influence on the capacity of joints by carrying out embedment tests, where the embedment strength parallel to the grain increased with increasing surface roughness. They could show that this increase is due to better stress distributions in the case of fasteners with rough surface; i.e. fasteners with rough surfaces activate shear, with less tensile stress peaks perpendicular to the grain directly underneath the fastener[28]. Numerical embedment models can then be used to estimate coefficients of friction, which Sjödin et al. called "fictitious parameters", as these cannot be directly evaluated from the tests. Estimated coefficients depend on the deformation level at which experimental and numerical curves are compared. Going back to the rope effect due to a normal force in the fastener axis, a straightforward test setup to directly determine coefficients of friction is to pull a smooth fastener from a timber piece. Schweigler et al. (2021) presented such tests where a smooth dowel was inserted into a predrilled timber piece, which was cut in two in order to exert compression perpendicular to the grain. Then, the smooth dowel is pulled out, and a direct

28 Such stress peaks appear already at small embedment or joint deformation as could be shown by Sandhaas (2012).

coefficient of friction could be derived. However, this test setup is not realistic as during loading, the fastener is more and more embedded in the timber, increasing friction. Therefore, Lemaître et al. (2023) extended the test programme and carried out biaxial tests, combining embedment and pull-out tests. They could show that friction coefficients (when pulling out the fasteners) increase with increasing embedment deformations. To conclude, it is clear that friction in all its forms is an important contributor to a joint's stiffness and capacity, but the challenge remains as to how it can be considered in design.

2.5 Overview over collected data

2.5.1 Data stemming from certification testing

The assembled databases contain test results of single fasteners (staples, nails, self-tapping screws) that are determined during certification procedures, which, in Europe, are carried out in accordance with EN 14592 (2012)[29]. Alternatively, European Technical Assessments (ETA) can be released that regulate production, properties und use of fasteners, where so-called European Assessment Documents (EAD) define the testing programme (e.g. EAD 130118-01-0603 (2019) for screws). Within this certification testing (EN 14592 or EAD), all or a part of the input parameters for design in accordance with the *EYM* are experimentally determined. The ***Timber Structures and Building Construction*** section at ***KIT*** started certification work already in the 70ies, and has thus many years of experience, also concerning the evolution of certification processes, and a very comprehensive database at its disposal. The oldest considered certification reports date back to 1997 and it is deemed useless to consider older results as the then used fasteners may not be representative of modern ones. Depending on the fastener type, a combination of the following parameters is experimentally determined and hence part of the databases, together with the geometrical data of the fasteners that must be measured in the framework of certification testing, see Figure 3-1 for staples, Figure 4-1 for nails and Figure 5-1 for screws.

[29] Or any other version of EN 14592 valid at the time of testing.

Tests with fastener only

- Tensile strength of wire f_u
  The tensile strength in [MPa] is derived from the tensile capacity, i.e. the maximum capacity of the wire during a tensile test, using the diameter d_{wire} of the wire. Wires represent the base material of staples, nails and screws. This property is only determined for staples and nails.
- Tensile capacity of fastener F_t
  The maximum value in [kN] is considered.
- Yield moment of fastener M_y
  A four-point bending test in accordance with EN 409 (2009) is carried out, and the yield moment in [Nm] is the ***maximum value taken until a bending angle of 45°*** or at the maximum bending angle before rupture of the fasteners.
- Torsional moment capacity M_{tor}
  The maximum value in [Nm] is considered. Hence, it is the value at fastener failure and consequently, M_{tor} is indeed a "moment capacity" in contrast to the yield moment M_y, where the ultimate capacity may not have been reached. M_{tor} is only determined for screws.

Tests with fastener and timber

- Withdrawal capacity of fastener F_w
  The maximum value in [kN] reached during a withdrawal test is considered. Fasteners were inserted at least 24h before testing and different timber pieces were used within a test series. Timber width was never recorded, but usually was 10 times the nominal diameter. The fastener tips were embedded in the timber.
- Head pull-through capacity of fastener F_{head}
  The maximum value in [kN] reached before or at a displacement of 15 mm of the test machine's crosshead is considered. In general and if not noted otherwise, the thickness of the timber pieces usually was 8 times the nominal diameter, whereas the thickness of the wood-based panels is the panel thickness. The timber width was never recorded, but usually was 10 times the nominal diameter. Different timber pieces were used within a test series.
- Insertion moment of fastener M_{insert}
  The maximum value in [Nm] reached until the screw head touches the timber is considered. Speed is always 100 rpm. This property is only determined for screws. The timber width usually was 10 times the nominal diameter. The screw tip always remained embedded in the timber.

2.5.2 Note concerning used wood and wood products

For all tests with fastener and timber, the density of the timber piece was measured, but must be considered being a gross density as usually, the timber piece was used for more than one test[30]. No information is available concerning the wood characteristics directly around the fastener, which can be expected to govern during the tests. For example, the local density might be lower or higher than the gross density, or local growth characteristics such as large annual rings, knots or small fibre deviations impacting on test results may be present. These characteristics do not need to be recorded during certification testing, and, looking at practical relevance, they are not required as the ***variety of materials on site will be even larger*** than ever present in a testing laboratory. This lack of local data, however, impacts on statistical evaluations as influencing factors are unknown. Moreover, it was not recorded if solid timber or glued laminated timber was used, which means that these two products are not differentiated. Furthermore, the timber pieces were stored at a normal climate of 20° C and a relative humidity of 65%. However, the moisture content was not measured. This is important to note as different wood and wood products were used. Above all, laminated veneer lumber made of beech (beech LVL) remains at very low moisture contents of about 6% to 8%, even when stored at normal climate over a long period, whereas spruce solid timber usually displays a moisture content between 11% and 12% when stored at normal climate.

Timber sizes usually were not recorded (see bullet points above). However, for all tests with fastener and timber, the timber sizes were always chosen such that only desired failures occurred. This means that no tests are included in the database where timber or fastener failures occurred instead; i.e. splitting of timber or fastener failures instead of withdrawal, head pull-through or full insertion.

2.5.3 Note concerning steel properties

Apart from their shape, staples, nails and screws have their basic material in common, which is steel. In that sense, they are comparable to fasteners used in steel structures, which, for dowels and bolts, is even more so the case as these two dowel-type fasteners exist in specific steel grades (e.g. "S235" or "8.8")[31]. In steel structures, the material properties used in design are (usually) yield strength f_y and tensile strength f_t. In particular for hardened steels with no distinct yield strength, the strength at 0.2% offset $R_{p0.2}$ is determined[32]. Indeed,

30 Except for M_{insert}, all tests including timber were always carried out using different timber pieces within one series. This means that with "timber piece used for more than one test", tests for different series are intended (e.g. one value of F_w of screws A, B, C was determined using the same piece).

31 See section 2.1.

32 Here, both properties, f_y and $R_{p0.2}$, are used as synonyms.

these two values (f_t and f_y) are given for dowels and bolts used in timber engineering. Nevertheless, considerable overstrength exists, and real strength values will often be much higher than nominal values. For instance in Sandhaas (2012), 12 mm steel dowels of grade S235 were ordered, which means that the tensile strength should have been f_u = 360 MPa. Tensile tests revealed a real tensile strength of f_u = 640 MPa. This leads to completely wrong yield moments and subsequently completely wrong failure modes – which is a serious issue in seismic design respectively in any design case that relies on a certain ductility of the joint behaviour. As a consequence, Blass and Colling (2015) investigated both dowel and joint behaviour and proposed a new equation to calculate the yield moment of dowels and bolts.

For staples, nails and self-tapping timber screws instead, steel properties must be experimentally determined. Whereas for staples and nails, this may seem superfluous, it is mandatory for screws. Staples and nails are produced from wires of a certain steel grade, and, only work hardening during the production, from wire to staple or nail, influences the final steel properties[33]. This should make it possible to define classes similar to dowels and bolts ("S235", "8.8"). Screws made of carbon steel instead are additionally hardened, using different hardening procedures well described in Ringhofer (2017), which lead to a large range of steel properties where strength values are often not evenly distributed over the screw's cross-section. Furthermore, stainless steel screws exist, which – in case of austenitic stainless steel – cannot be hardened. Concerning tensile tests, only tensile capacities are determined, with the exception of fully threaded screws used as reinforcement perpendicular to the grain. For these, also $R_{p0.2}$-values are evaluated, using the inner diameter of the screws to transform capacity into strength[34]. For design of timber joints, also the yield moment is needed and experimentally determined. For steel engineers, the terminology "yield moment" used in timber engineering is misleading, because it implies that the elastic bending capacity of a round section is meant, where the outer fibres just reached the yield strength f_y. For timber engineers instead, the yield moment is a property of a dowel-type fastener that includes plastic deformations, but where it is unclear how far the fastener's cross-section has plastified (see also section 5.3.2).

For comparison and analyses of steel property data, experimental values must be normalised in order to compare direct test data (M_y, F_t, M_{tor}) between fasteners with different diameters. This is straightforward when calculating tensile strength from tensile capacity, where the only debatable issue is which diameter should be considered. To normalise the yield moment M_y, Eq. (3-2) could be used, the equation to calculate the full plastic bending capacity of a round section, resulting in Eq. (3-1). However, the value for M_y depends on the bending angle and thus on a deformation, with no information on the plastification. Furthermore, work hardening effects will occur during testing and as a consequence, the transformation in accordance to Eq. (3-1) does not deliver a "yield strength f_y" for staples, nails and

33 In general. However, also hardened nails exist that undergo a hardening procedure analogously to screws.

34 This is already a modelling step, as a screw's cross-section in the threaded area is not round (Ringhofer, 2017).

screws. Nevertheless, the use of Eq. (3-1) can be motivated considering Figure 2-3 on the left, which shows three exemplary results of M_y-tests that can be classified in three classes F1 to F3 depending on the general shape of the moment-bending angle curves (Steilner et al., 2022). Classes F2 and F3 are typical curves for hardened self-tapping screws[35], where the difference is an earlier rupture of a class F3 screw. For fasteners following these curves, the evaluated M_y-value does not depend on the bending angle, because a maximum value is reached before a bending angle of 45° (and after an angle of 20°), and the decreasing curve after this maximum value indicates that full plastic bending capacity is reached, which is not overlapped by work hardening effects. Class F1 behaviour can be observed for staples, nails[36] and (stainless steel and unhardened carbon steel) screws, and the increasing slope makes interpretations difficult. Here, the evaluated M_y-value depends on the bending angle at which it is determined, with a difference of 5% between $\alpha = 20°$ and $\alpha = 45°$. Considering the large bending angles[37] and the difference of about 5%, however, also here the assumption of full plastification is acceptable. The last point to be discussed is the difference between $R_{p0.2}$-values and f_t-values, as only the latter are known and can hence be used in design. In Figure 2-3 on the right, three exemplary stress-strain curves measured during tensile tests on a mild steel, a dowel labelled grade S235 and a fully threaded screw made of hardened carbon steel are shown. Neither the dowel nor the very high strength steel screw show a distinct yield strength, and the difference between $R_{p0.2}$-values and f_t-values is 6%. Work hardening effects are not strong, not even for the dowel, indicated by the shallow shape of the curve in the plastic range before necking starts. The curve shapes are indeed in stark contrast to the curve for the mild steel with a clear yield strength and strong work hardening before necking.

Steel properties of the investigated fastener types are hence complex due to hardening of fasteners (usually only screws) and work hardening effects during production (staples, nails, screws). Moreover, geometries of screws and non-smooth shank nails are complicated. In particular concerning the yield moment, the values contained in the databases could be considered being random, as they depend on the fastener's bending angle. Strictly speaking, no "yield strength f_y" is calculated for staples, nails and screws using Eq. (3-1). Therefore, the stress value calculated in accordance with Eq. (3-1) is called "yield moment stress σ_{My}" in this study.

35 Referring to the caption of Figure 2-3, not only the curves are exemplary, but also the assignment of fasteners to the classes F1, F2 and F3. The discussion highlights general trends, and exceptions exist.

36 A general difference can be observed between nails made of carbon and stainless steel, with nails made of carbon steel showing a sharper bend in the transition area between elastic and plastic behaviour.

37 And the discussion in Sandhaas (2012), where it is shown that full plastic bending capacity can be reached at small bending angles, depending on the curvature and neglecting work hardening effects.

To be as clear as possible when addressing different steel properties, in particular those derived from direct test data (i.e. strength values), the following notation is consistently used throughout this study:

- A capital ***F*** to denominate capacity values in N.
- A lowercase ***f*** to denominate strength values in MPa.
- The Greek letter σ to denominate stress values in MPa.
- Indexes are used to clarify all further points; e.g. which diameters are applied to transform capacity into strength values.

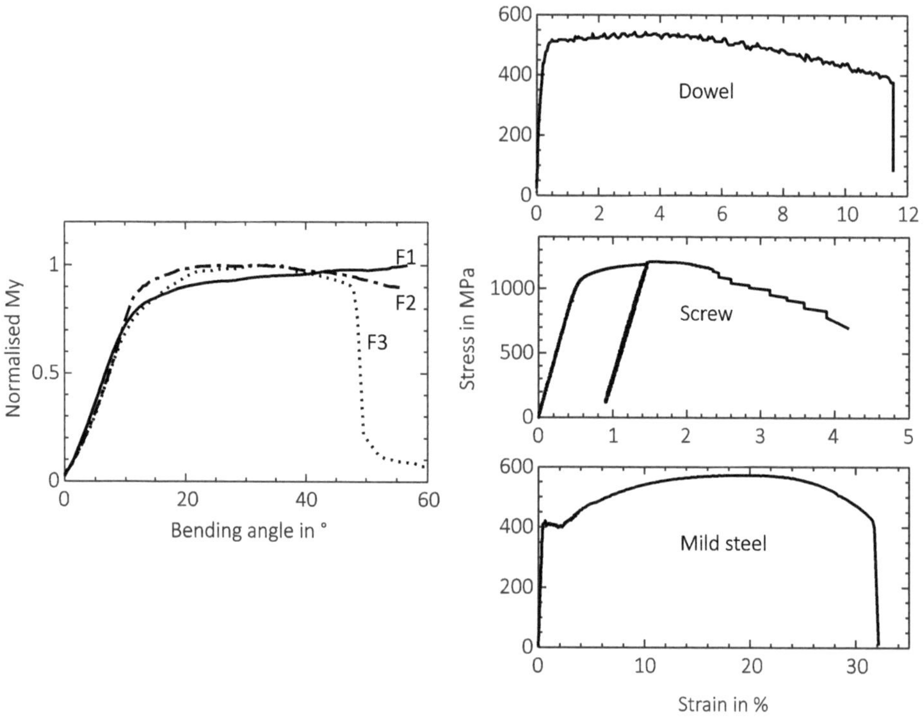

Figure 2-3 Left: Three typical moment-bending angle curves of self-tapping screws (Steilner et al., 2022). Right: Typical stress-strain curves; top: 12 mm dowel labelled grade S235, $R_{p0.2}$ = 510 MPa, f_t = 544 MPa; centre: 6 mm fully threaded screw, d_i = 2.9 mm, $R_{p0.2}$ = 1133 MPa, f_t = 1209 MPa; bottom; mild steel, $R_{p0.2}$ = 416 MPa, f_t = 573 MPa. The shown curves are **exemplary** curves, showing general trends, and exceptions exist.

2.5.4 Note concerning data quality

Each fastener type can be described by its geometry and its material properties. Not all of these data is available, however, as the databases were assembled retrospectively, and not everything was recorded at the time of testing as it was not required by EN 14592 (2012) or the EAD (2019). Such unrecorded data for instance encompass thread flank angles of self-tapping screws or information on coating. In addition, certain data may be missing in some reports, for instance the nail tip length was not always recorded. Particular challenges arise when looking at e.g. available tip types of screws, which represent a large variety and are difficult to classify (see Figure 5-3). Also production issues such as the effect of worn tools or manufacture of very long screw threads, which are rolled in sections and hence small notches may be present, are not captured. This is a critical point, because complete and unequivocal datasets are needed for proper regression analyses, and the databases are lacking exactly this. Influencing factors on mechanical properties of fasteners will exist that were not recorded and which, therefore, cannot be identified. This statement refers directly to opportunity 4 of section 1.1, without having carried out any analysis.

2.5.5 Note concerning test protocols

Certification test protocols do not exhaustively describe how the tests must be carried out. This has various implications. For instance, head pull-through parameters evaluated at different testing bodies are difficult to compare as the relevant standard EN 1383 (2016) does not specify at which deformation the head pull-through capacity should be determined. Furthermore, the standards may be rather unspecific as to how to select the timber pieces for withdrawal (EN 1382, 2016) and head pull-through tests (EN 1383, 2016). Both standards state that timber pieces shall be selected in accordance with EN ISO 8970 (2010), which aims at "getting specimens with a uniform density comparable with the mean density of the timber to which the test results should be applied." EN ISO 8970 does however not explicitly state that different timber pieces should be selected for testing and that, e.g., the withdrawal capacity should not be evaluated using only two timber pieces (which may fulfil the requirements of EN ISO 8970), see also footnote 30. prEN 14592 (2017) gives more precise indications on how to account for different coefficients of variation of the selected test specimens and the timber grades the test specimens are applied to. Another example are issues around EN 409 (2009), where bending tests to determine the yield moment are specified. Again, the standard does not describe in detail how tests must be carried out (for this, see section 5.3.2). In particular, all deformation-dependent values are critical and prone to measuring and interpretation errors, and also this statement directly tackles opportunity 4.

2.5.6 Note on comparability of data

When data stemming from different test series carried out in different periods, by different people and/or which were not part of a single bespoke testing programme, the first and foremost question to be answered is whether the assembled data can be compared with each other or not. This is the case here, although the oldest datasets stem from 1997[38] and the newest from 2019. There are four main reasons for this:

- Only a certain type of data was assembled that encompass geometrical and experimental values that are rather straightforward and easy to determine[39] and that are evaluated using the same testing principles during the years.
- The scope of all considered tests remained the same throughout the years; i.e. evaluating data for the calculation of characteristic input parameters for joint design.
- The tests were all carried out at KIT, during a period with no change in people defining internal technical guidelines. So, at least, all tests were carried out at the same institution, by only few different people and using only few different test setups and machines.
- Test standards did change between 1997 and 2019, but general procedures at KIT did not change and neither did the accurateness and carefulness of execution change. For instance, only withdrawal data is given where withdrawal failure occurred; i.e. failure modes were never mixed, and always only the sought-after parameter was established and not something else.

38 Staples and nails; the oldest data for screws are from 2010.

39 Remembering the previous notes, the assembled values were not as "straightforward and easy to determine" as stated here. Nevertheless, e.g. withdrawal tests are considerably less complex than e.g. joint tests. Even less complex are the tests determining F_t and M_{tor}. The least susceptible data are the geometrical data.

3 Staples

3.1 General

The first database presented in this study contains data for staples. This database is an extended version of the database collected by de Proft (2017), and it is the most limited of all assembled databases, as only few tests with staples were carried out in the last decennia, and the latest available tests date back to 2013. Obviously, a limited database will lead to limited analyses. Nevertheless, information on scatter of results and experimental gaps can be gathered. The lack of testing data is reflected in a lack of research in general. One exception to this is a publication from 1973 (Möhler et al., 1973), whereas more recent publications focus on the cyclic behaviour of stapled joints (Sartori and Tomasi, 2013). Triggered by the lack of data, a recently finished research project dealt with minimum distances and spacing as well as load-carrying capacities of stapled joints (CEN/TC 250/SC 5/WG 5 N 169, 2021). A preliminary study with staples in hardwood products instead was carried out by Vedovelli (2020).

Concerning design, current approaches for stapled joints are based on the European Yield Model (*EYM*), where one staple leg is considered as a nail. Usually, staples are inserted with an angle between staple crown and fibre direction. In current building practice, staples are generally used to fasten sheathing, e.g. an OSB or a particleboard, to a (softwood) timber frame substructure. As staples are driven in using pneumatic machines, the production of stapled joints occurs mostly in factories and a high degree of automation is possible. The insertion quality is influenced by the chosen pressure, which defines the insertion depth of the staple crowns. Additionally, the slender staple legs buckle easily. Although staples are versatile fasteners leading to efficient, ductile and cost-effective joints, their use is not widespread in all European countries. Moreover, staples were not significantly modified for decades in terms of their production or used materials. This can, at least partly, explain the lack of publications and experimental results on both staple properties and stapled joints.

3.2 Database

In accordance with EN 14592 (2012), characteristic values for the yield moment, the withdrawal and head pull-through parameter must be declared. Furthermore, EN 14592 requires control of the geometric properties and evaluation of the tensile strength of the wire, which must be higher than 800 MPa. The assembled database presented here contains results from 34 reports and the following individual test values:

- 352 tensile strength values in 89 series[40] of the wire f_u, of which 103 tests in 29 series on staples made of stainless steel
- 260 yield moments M_y in 29 series, of which 90 tests in 11 series on staples made of stainless steel; the tests were carried out about the weak axis of the staple leg.
- 1603 withdrawal capacities F_w in 84 series, of which 389 tests in 32 series on staples made of stainless steel
- 120 head pull-through capacities F_{head} in 6 series, no staples made of stainless steel

Staples from ten different companies are included and the reports date from 1997 until 2013 where, however, 80% of the tests results were evaluated between 2009 and 1997. Both nominal and measured geometrical properties are included (as far as recorded), definitions see Figure 3-1, and distinction was made between galvanised carbon steel and stainless steel. Properties of staples made from galvanised carbon steel are not different to those of staples made from stainless steel; see e.g. Figure 3-2. The diameter d_{wire} of the wire corresponds to the nominal diameter d of the staple leg. The staple itself has a bevelled cross-section with different thickness a and width b, see Figure 3-1. The final shape of the staple is a result of cold forming processes during which the straight wire with its round cross-section is transformed into a staple with a bevelled cross-section, a crown and two legs. Nominal diameters d ranged from 1.51 mm to 2.00 mm. The staples were partially or fully resin-coated.

[40] This terminology will be used consistently throughout this study. A "series" comprises a certain quantity of individual test results; i.e. the number of series and not the number of individual test results is crucial to understand the diversity of fasteners resp. tests contained in the databases.

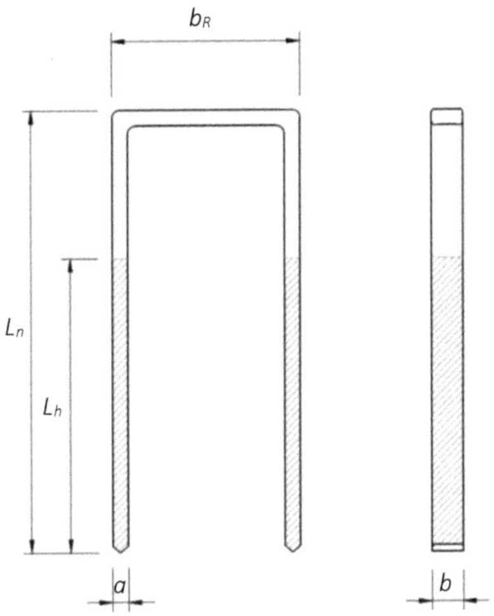

Figure 3-1 Geometry of a staple, with thickness a and width b of staple legs, width of the crown b_r, length of the staple L_n and length of resin-coated parts L_h. The staples' nominal diameter d corresponds to the wire diameter d_{wire}.

3.3 Tensile strength and yield moment

Tensile strength

In total, 352 individual results of tensile tests on wires, grouped in 89 test series with two to ten tests per series, and 260 individual results of yield moment tests on staple legs, grouped in 29 test series with five to 20 tests per series, are available. The final bevelled shape of the staple means that the round wire must be passed through a die in order to produce the staples. This cold forming of the wire changes the staple's properties through work hardening effects, and the experimentally determined tensile strength of the wire, before cold forming, will be lower than the tensile strength of the staple after cold forming[41]. This effect cannot be investigated as no tensile tests on finished staples were carried out. The statement, however, can be underlined considering Figure 3-2 on the left where the mean tensile strength of the wire is lower than the mean "yield moment stress σ_{My}" of the staples produced with the tested wire. σ_{My} is calculated in accordance with Eq. (3-1), which is based on the equation to calculate the full plastic bending capacity M_{pl} of a circular cross-section, Eq. (3-2). In Eq. (3-1), the wire diameter d is considered, as this is the nominal diameter of a staple.

[41] It can be even assumed that the steel properties will be different within a staple itself, with higher properties close to the area of the 90 degrees angle between crown and legs, as the bending of the wire is a further cold forming process.

$$\sigma_{My} = \frac{6 \cdot M_y}{d^3} \tag{3-1}$$

where

σ_{My} "Yield moment stress" in MPa, per leg

M_y Yield moment in Nmm, per leg

d Nominal diameter of staple in mm

$$M_{pl} = \frac{1}{6} \cdot f_y \cdot d^3 \tag{3-2}$$

where

M_{pl} Full plastic bending capacity

f_y Yield strength

d Diameter

However, tensile strength and "yield moment stress" of the staple cannot be compared unconditionally. Firstly, the determination of the yield moment includes bending of staple legs up to 45°. Such a large bending will lead to work hardening effects and hence to an increase in tensile strength. Considering that staples are made from ductile steel – legs must be bent by 90° without any rupture or cracks – work hardening effects will be rather important (contrarily to very high strength steel). Secondly, the yield moment is determined by bending the staple leg about its weak axis a, where a is smaller than the nominal diameter d used in Eq. (3-1). Consequently, the "yield moment stress" will be even higher when considering a instead of d. Finally, both properties f_u and M_y are determined using different material, often even from different batches.

In Figure 3-2 on the left, one outlier belonging to tensile tests on 1.83 mm wires and staples is observed (where the mean yield moment strength is lower than the tensile strength of the wire), but cannot be explained conclusively. The most obvious hypothesis is that the delivered wires used for tensile tests did not correspond to the wires used to produce the staples for the yield moment tests. Another possible hypothesis could be that the galvanisation of the staples, which represents a heat treatment, reduced the strength that was determined on the non-galvanised wire.

Figure 3-2 on the right shows the tensile strength values of all wires versus the nominal diameter (both wire and staples produced thereof). Not all wires reached the mandatory strength of 800 MPa[42] (and were hence rejected) and no difference can be seen between wires made of stainless or carbon steel. The linear trendline[43] shown in Figure 3-2 on the right, including all test results except for the values lower than 800 MPa, shows a general

[42] Eurocode 5 (2010) requires a minimum tensile strength of 800 MPa of wires, see section 8.4 clause (6).

trend of decreasing tensile strength with increasing leg diameter. Considering the scatter of test results, this trend is rather weak and should not be overestimated. It however reflects more valid observations made for nails, where repetitious drawing passes of smaller diameter nails leads to a work hardening effect with subsequent higher tensile strength in comparison to larger diameter nails with less drawing passes (see section 4.3). To conclude, tensile tests on wires, as regulated currently, have no other purpose than to control the steel quality of the raw material – which, seeing the tensile strength values < 800 MPa in Figure 3-2 on the right, must remain mandatory tests – and do not give any meaningful information on tensile properties of staples.

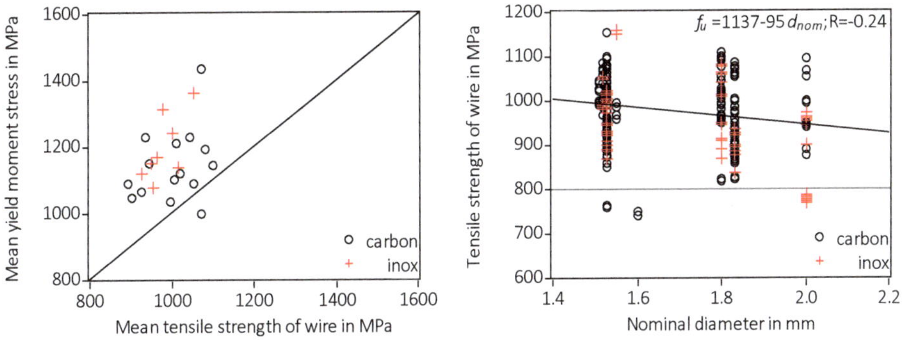

Figure 3-2 Left: Mean tensile strength of the wire f_u versus mean yield moment stress σ_{My} in accordance with Eq. (3-1). 23 different series with 250 individual values for M_y and 119 for f_u. Lowest value of σ_{My} is 997 MPa. Right: Tensile strength of the wire f_u versus nominal diameter d, 352 individual test results. The shown (linearly decreasing) trendline[43] includes all results except values lower than 800 MPa.

[43] Within this study, **all trendlines are extrapolated** and range over the whole bandwidth of the horizontal axis.

Yield moment

Amendment A2 of Eurocode 5 (EN 1995 1-1/A2, 2014) gives Eq. (3-3) to calculate the characteristic value for the yield moment of staples:

$$M_{y,Rk} = 150 \cdot d^3 \tag{3-3}$$

where

$M_{y,Rk}$ Characteristic value for the yield moment in Nmm, per leg

d Nominal diameter of staple in mm

Eq. (3-3) is based on Eq. (3-2) with f_y = 900 MPa. Considering the mean value of the yield moment strength σ_{My} given in Figure 3-2 on the left, this value of 900 MPa holds, even though Eq. (3-3) determines the ***characteristic*** value of the yield moment (see also Figure 3-3 on the left, where min $\sigma_{My,k}$ = 970 MPa). The original equation given in Eurocode 5 (EN 1995 1-1, 2010), $M_{y,Rk} = 240 \cdot d^{2.6}$, was based on reflections concerning the observed decrease of tensile strength with increasing diameter (Figure 3-2 on the right). The original equation is analogous to the equation for nails, for more details see section 4.1. It must be kept in mind that during testing, staples will be bent around their weak axis and in practice, the bending may occur at any angle depending on the angle between orientation of the staple legs and load direction.

Apart from Eq. (3-3), characteristic values for the yield moment of staples can also be taken from Declarations of Performance (DoP). Figure 3-3 on the left shows the yield moment versus the nominal diameter of the staple leg, including all experimental values (the black circles), declared characteristic values (the red crosses) and a curve representing Eq. (3-3). Furthermore, the characteristic values per test series calculated in accordance with EN 14358 (2016) are given, the green triangles, as these would be the declared characteristic yield moments of the tested staples. The characteristic values are calculated assuming a lognormal distribution and using a fixed standard deviation of 0.05. The observed standard deviation is shown in Figure 3-3 on the right and is always smaller than 0.05 in all test series. The Eurocode 5 prediction, Eq. (3-3), is still more conservative.

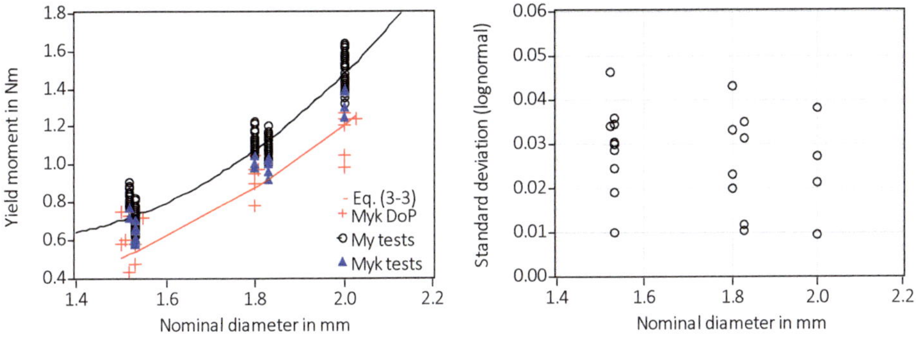

Figure 3-3 Left: Black circles: Experimentally derived yield moments (260 values in 25 test series). Red crosses: Declared characteristic values for the yield moment (values from DoP). Green triangles: Characteristic values calculated acc. to EN 14358 (2016); lowest $M_{y,k,tests}$ = 0.58 MPa leading to $\sigma_{My,k}$ = 970 MPa. The black line shows the quadratic trendline of all test results (black circles) and the red line shows Eurocode 5 predictions, Eq. (3-3). Right: Standard deviation of experimental values of M_y based on lognormal distribution.

Characteristic values in accordance with Eq. (3-3) lie clearly below all experimental values. This is astonishing considering that in Eq. (3-3), the implicitly included steel strength is with 900 MPa close to the mean tensile strength of 967 MPa. Work hardening effects caused by the bending of the staple legs during the determination of the yield moment may explain this, as these hardening effects may be observed to a lesser extent during tensile tests. Some declared characteristic values taken from DoP are lower than the values calculated with Eq. (3-3) and some of them are rather high compared with experimental values. These high values are possible considering the scatter in steel grades. Figure 3-2 on the right shows tensile strength values ranging from 740 MPa to 1160 MPa and staples made of the latter steel will have high characteristic values for the yield moment. Generally, Eq. (3-3) seems to be rather conservative, with all experimental values being clearly higher. Having only 260 individual test values at disposal, Eq. (3-3) is however deemed sufficient and no changes to Eurocode 5 are needed. A reformulation of Eq. (3-3) could be included, however, in order to illustrate the mechanical background:

$$M_{y,Rk} = \frac{1}{6} \cdot f_{y,k} \cdot d^3 \tag{3-4}$$

where

$M_{y,Rk}$	Characteristic value for the yield moment in Nmm, per leg
$f_{y,k}$	Characteristic yield strength[44] in MPa, $f_{y,k}$ = 900 MPa
d	Nominal diameter of staple in mm

44 It can be discussed if "yield strength" is the proper terminology here, see also section 2.5.3.

3.4 Withdrawal

In total, 1603 individual test results are available, grouped in 72 test series. 768 of the tests were withdrawal tests in radial direction, 766 in tangential direction, and the direction of withdrawal is unknown for 69 tests. Together with the withdrawal capacity F_w in kN (= maximum machine load), also the density ρ in kg/m^3 was measured, where only the mean density was given for 65% of the test results. No density was given for 3.7% of the test results, and in that case, a mean density of 420 kg/m^3 was assumed. It is hence meaningless to show a histogram of the density to understand the scatter. The recorded density was measured as global value of one timber piece. The used timber was stored in a normal climate with 20°C and 65% relative humidity. The moisture content was not measured. The used wood species was spruce (*Picea abies*). All staples were resin-coated and the penetration lengths L_{ef} ranged from $10 \cdot d$ to $22 \cdot d$. The staples were inserted directly into the timber; i.e. no wood-based panel or other intermediate layer was used.

For most test results, no indication is given concerning the angle θ between staple crown and fibre direction. If a value was given, then $\theta = 90°$. It is very probable however, that all withdrawal tests were carried out at $\theta = 90°$, as no distinction is made within test results, which would have been the case if various angles were chosen[45].

In the current Eurocode 5 (EN 1995 1-1, 2010), capacity to axial loading of staples is calculated analogously to nails, see Eq. (4-2), and characteristic withdrawal parameters $f_{w,k}$ and head pull-through parameters $f_{head,k}$ are declared in DoP. The database contains only withdrawal capacities, which, before executing any analysis, must be normalised by applying Eq. (3-5)[46].

$$f_w = \frac{F_w}{2 \cdot L_{ef} \cdot d} \tag{3-5}$$

where

f_w	Withdrawal parameter in MPa, per leg (hence, factor 2)
F_w	Withdrawal capacity in N
L_{ef}	Penetration length in mm
d	Nominal diameter of staple in mm

[45] Standards changed; in the 1999 version of EN 1382 for instance, the angle between staple crown and fibre direction was 0° and 90°, whereas in the 2016 version, withdrawal tests with staples should be carried out with $\theta = 30°$.

[46] This equation uses the projected area of the staple legs and not their surface area. This is currently the conventional method in accordance to EN 1382 (2016).

Figure 3-4 shows all 1603 individual withdrawal parameters versus density, where lower values were determined when withdrawing from the tangential direction. This is due to less homogenisation in comparison to withdrawal from the radial direction as then, the legs are inserted orthogonally to the annual rings and hence penetrate both early- and latewood and various annual rings. Generally, the dependency of the withdrawal parameter f_w on the density ρ is too weak to derive an equation where f_w solely depends on ρ.

To facilitate design, a constant characteristic withdrawal parameter per staple leg could be given. The horizontal line in Figure 3-4 illustrates this; the line at 6.9 MPa indicates the observed 5th percentile of all 1603 withdrawal test results. If considering only the f_w-values determined using spruce with densities of 390 kg/m³ < ρ < 450 kg/m³, the observed 5th percentile remains the same with 6.8 MPa. A constant characteristic value for the withdrawal parameter per leg of resin-coated staples could hence be 7 MPa. Declared characteristic values range from 4.8 MPa to 7.3 MPa, where 23 of 29 (80%) $f_{w,k}$-values taken from DoP are lower than 6.8 MPa. It is up to code-writing committees to decide if a constant characteristic value should be lower than 7 MPa.

One remark must be made here concerning the evaluated withdrawal parameters. These are derived from tests where staples are directly inserted into the timber. This, however, is not relevant in practice as staples are used to fasten wood-based panels onto timber. That means that staples are driven through wood-based panels before penetrating timber, which may lead to an abrasion of the resin coating with subsequent lower withdrawal resistance. The thicker the wood-based panel is, the less resin coating may remain and hence the lower the withdrawal resistance may be.

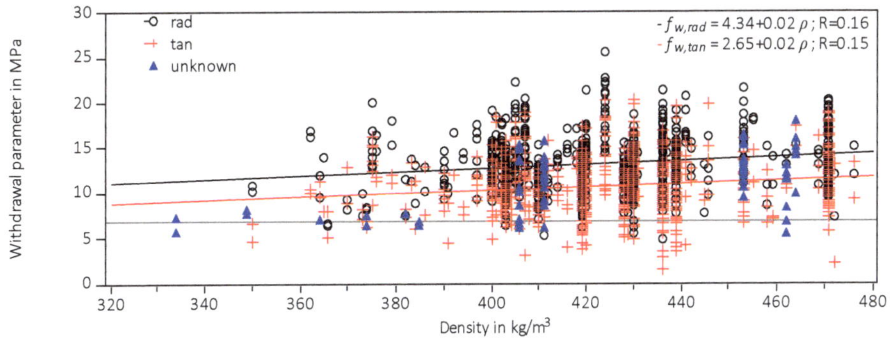

Figure 3-4 Individual withdrawal parameters versus density; differentiation with respect to withdrawal direction, 1603 individual values. Trendlines for radial (black) and tangential (red) values are shown. The horizontal line at 6.9 MPa corresponds to the observed 5th percentile considering all f_w-values.

3.5 Head pull-through

In the database, only 120 head pull-through capacities derived for staples from one producer are available; 60 values for staples with $d = 1.53$ mm, 20 values for staples with $d = 1.80$ mm and 40 values for staples with $d = 2.00$ mm. Concerning density and moisture content, the same procedure was used as for the withdrawal tests. The tested angles between staple crown and fibre direction were 0° and 90°, and only spruce was used, where the thickness of the timber was not specified. The evaluated head pull-through capacity F_{head} [47] is normalised as follows:

$$f_{head} = \frac{F_{head}}{b_r \cdot b} \tag{3-6}$$

where

f_{head}	Head pull-through parameter in MPa
F_{head}	Head pull-through capacity in N
b_r	Width of staple crown in mm, see Figure 3-1
b	Width of staple leg in mm, see Figure 3-1

Figure 3-5 shows all 120 test data versus the density where the difference in head pull-through parameter for angles of 0° and 90° between staple crown and fibre direction can be seen. Moreover, the large scatter with no clear trend of higher f_{head}-values for higher densities is evident (the density range is however rather small with $338 \leq \rho \leq 398$ kg/m^3). The horizontal lines indicate the observed 5th percentiles, evaluated for 60 test results per angle. These 5th percentiles could be used as lower boundary values where, however, more test should be carried out before deciding on any constant head pull-through parameter. Moreover, the results contained in the database are valid only for head pull-through using solid softwood of a certain thickness. The most relevant case for practice, staples to fasten a wood-based panel to a timber frame substructure (i.e. head pull-through through wood-based panels at $\theta = 30°$), is not covered, and will yield lower values due to the in general reduced thickness of the panels.

[47] Due to the age of the test reports, it is not clear at which deformation F_{head} was read, see also section 5.6.3.

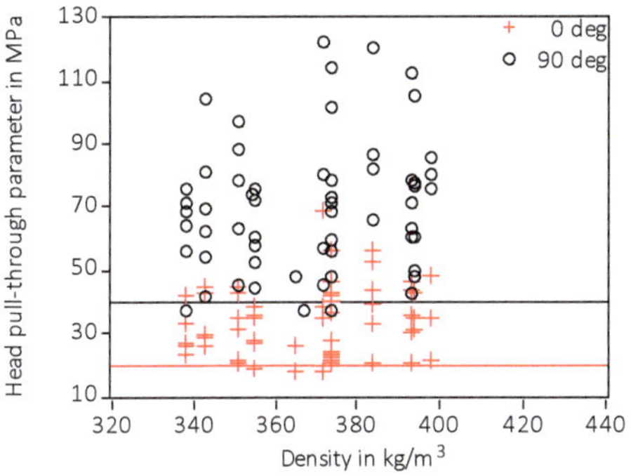

Figure 3-5 Individual head pull-through parameters versus density, differentiation with respect to angle between staple crown and fibre direction, 60 tests at 0° and 90° respectively. The two horizontal lines at 39.7 MPa and 19.2 MPa indicate the observed 5th percentiles.

3.6 Conclusions

In general, it can be said that the assembled database must be extended in order to allow for more thorough and relevant investigations. This encompasses tests that determine the steel properties of current staples or head pull-through tests using wood-based panels. Still, conclusions addressing the input parameters for the *EYM*[48] can be given. Concerning the yield moment, current regulations in Eurocode 5 are sufficient. Concerning withdrawal and head pull-through parameters, the observed large scatter does not allow for derivation of equations with which characteristic values can be calculated; at least not without systematic investigations into the relationship between parameters and density. Conservative lower bound values sufficient for most design cases are possible and pragmatic.

48 A persisting problem not discussed here remains the embedment strength, as current test methods, i.e. the full hole tests in accordance with EN 383 (2007), are not suited for staples with their small diameters.

4 Nails

Note: Parts of this chapter were already published in Sandhaas and Görlacher (2018).

4.1 State-of-the-art

Joint design in the current Eurocode 5 (EN 1995 1-1, 2010) is based on Johansen's model (1949) that firstly was applied to nailed joints by Moeller (1951). Since then, considerable research was put into further development of methods to establish the ultimate characteristic load and deformation behaviour as discussed in Ehlbeck (1979), who gives a concise and comprehensive summary of the state of the art in the late seventies. Ehlbeck already discussed input parameters necessary for the design of nailed joints such as embedment strength and yield moment as well as the contribution of the rope effect to the joint capacity and hence the withdrawal performance of non-smooth shank nails. The background discussed by Ehlbeck is still representative today as research concerning joints with dowels, bolts and screws were and still are in the focus whereas nailed joints are less represented in current research. An exception to this is the work done by Whale and Smith in the eighties concerning embedment strength (Whale and Smith, 1986a, 1986b) and investigations by Blaß in the early nineties concerning group effects in nailed joints (Blaß, 1990, 1991). Nails are driven in using a hammer or pneumatic machines, where in particular the small diameter nails are usually available in coils or strips (plastic or paper) that are inserted in the appropriate machines. In comparison to staples, therefore, the insertion quality is not only influenced by the chosen pressure but also by human aptitude if inserted manually.

In accordance with EN 14592 (2012), characteristic values for the yield moment M_y, the tensile capacity F_t, the withdrawal and head pull-through parameters F_w and F_{head} must be declared, usually in individual DoPs of the producers. Furthermore, EN 14592 requires control of the geometric properties and evaluation of the tensile strength of the wire, which must be higher than 600 MPa. Not all characteristic values must be experimentally derived however. For smooth shank nails for instance, the characteristic yield moment $M_{y,Rk}$ can also be calculated.

For round, smooth shank nails, Eurocode 5 gives the following Eq. (4-1):

$$M_{y,Rk} = 0.3 \cdot f_u \cdot d^{2.6} \qquad (4\text{-}1)$$

where

$M_{y,Rk}$	Characteristic value for the yield moment in Nmm
f_u	Tensile strength of wire in MPa, where f_u = 600 MPa
d	Nominal diameter of nail in mm

Eq. (4-1) is based on work done by Werner and Siebert (1991) who proposed the empirical relationship $f_{y,k} = 1100 \cdot d^{-0.4}$. Their proposal is based on an observed increase of yield stress[49] with decreasing nail diameter although the nominal strength of the wire was assumed as 600 MPa. This leads to a yield moment $M_{y,Rk} = f_{y,k} \cdot d^3/6 = 1100 \cdot d^{-0.4} \cdot d^3/6 = 180 \cdot d^{2.6}$ which was implemented in ENV 1995 1-1 (1993). As $180 = 0.3 \cdot 600$, Werner and Siebert's equation was modified to be equal to Eq. (4-1) in order to give only one equation to calculate $M_{y,Rk}$ of nails, bolts and dowels[50]. This means that Eq. (4-1) is valid only when a wire tensile strength of 600 MPa is inserted. This value of 600 MPa is mandatory even if the actual value is higher, which is the case for diameters less than 4 mm. It is the exponent of 2.6 in Eq. (4-1) that reflects the observed increase of yield strength (up to 1000 MPa for 2 mm nails) with decreasing nail diameter, which can be explained with work hardening due to cold drawing.

Also for the parameters f_w and f_{head}, regression equations are given in Eurocode 5 for short term loaded smooth shank nails:

$$f_{w,k} = 20 \cdot 10^{-6} \cdot \rho_k^2 \quad \text{and} \quad f_{head,k} = 70 \cdot 10^{-6} \cdot \rho_k^2 \qquad (4\text{-}2)$$

where

$f_{w,k}$	Characteristic withdrawal parameter in MPa
$f_{head,k}$	Characteristic head pull-through parameter in MPa
ρ_k	Characteristic density in kg/m³

[49] Werner and Siebert derived a "yield stress f_y" based on experimentally derived yield moments in analogy to Eq. (3-1); Werner and Siebert's f_y hence corresponds to the " yield moment stress σ_{My}" used in this study (where, however, the used testing machine to determine M_y was different).

[50] The background of the equation for bolts and dowels is, however, completely different, see Blaß et al. (2000).

4.2 Database

The database consists of in total 9774 tests taken from 100 reports on mostly ringed shank nails (rings 77%, spiral nails with threads 5%) and wires (11%). Special ringed shank and spiral nails with smooth intermediate shanks as shown in Figure 4-1 on the left constituted 5.3% of the overall database. Smooth shank nails constituted 1% of the database and square nails only 0.3%. For smooth shank nails, only wire strength was tested and no other parameters are available. Consequently, nail types such as smooth shank nails or squared nails are consistently under-represented in the database. Available diameters range from 2.1 mm (3.2%) to 6.12 mm (0.1%). Most nails had diameters of 2.5 mm (11.2%), 2.8 mm (13.8%), 3.1 mm (13.7%), 4 mm (19.1%) and 6 mm (11.2%). The ratio of inner over outer diameter of non-smooth shank nails ranged between nearly 1.0 and 0.73. Nails from 37 different producers were included, and the tests were carried out between 1997 and 2015. Similar to staples, it is not considered useful to enlarge the database with older results as used steel grades and production technologies may have changed since then, and analyses including these old nails would not be representative for modern nails. The geometrical properties given in Figure 4-1 on the right are also generally recorded in the database. The number of tests per parameter is given in Table 4-1. As properties of nails made from stainless steel do not differ significantly from those of all other nails, see e.g. Figure 4-3 on the left, no difference is made in analyses.

Withdrawal capacity F_w and head pull-through capacity F_{head} were evaluated using mainly spruce (*Picea abies*). Few withdrawal results are available where laminated veneer lumber was used; only 40 tests with softwood LVL and 20 tests with beech LVL were carried out without predrilling. The used timber was stored in a normal climate with 20°C and 65% relative humidity. The moisture content was not measured, and the density must be considered being a gross density without information of the wood characteristics directly around the nails. All withdrawal and head pull-through tests were carried out with an angle α = 90° between nail axis and grain direction.

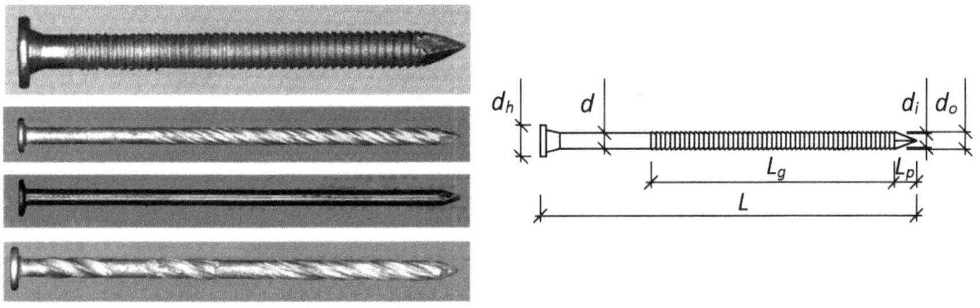

Figure 4-1 Left: Nail shapes in database. From top to bottom: ringed shank nail, spiral nail, smooth shank nail, special spiral nail. Right: Geometrical properties with d = nominal nail diameter, d_i = inner diameter, d_o = outer diameter, d_h = head diameter, L_g = length of non-smooth shank, L_p = tip length.

Table 4-1 Composition of database.

	Tensile strength f_u of wire	Yield moment M_y	Tensile capacity F_t of nail	Withdrawal capacity⁺ F_w	Head pull-through capacity F_{head}
Number of tests	1076 tests in 257 series	2854 tests in 353 series	1160 tests in 119 series	3644 tests in 186 series	1039 tests in 68 series
Of which stainless steel	203 tests in 55 series	369 tests in 46 series	195 tests in 22 series	390 tests in 19 series	60 tests in 4 series
Of which hdg*	-	265 tests in 32 series	178 tests in 15 series	320 tests in 21 series	220 tests in 15 series

* hdg = hot-dip galvanised

⁺ 60 tests were carried out using softwood LVL instead of solid softwood.

4.3 Tensile strength and tensile capacity

4.3.1 General

As already stated, Eq. (4-1) is based on work by Werner and Siebert (1991), who observed an increase of strength with decreasing diameter. The analysis of the current database confirms this finding. This is shown in Figure 4-2 where the nail strength f_t was calculated using the tensile capacity F_t and the nominal diameter d[51]. The increase of tensile strength with decreasing diameter of both wires and nails can be explained with work hardening due to cold drawing as multiple passes are needed to produce smaller wires and hence smaller diameter nails. The effect can be also seen on the level of individual test values, see Figure 4-3 on the left, where it is also shown that the tensile strength of stainless steel wires hardly differs from that of carbon steel wires[52]. The observed influence of the diameter on both wire and nail tensile strength is thus caused by production methods and is usually considered applying (empirical) regression equations; e.g. Eq. (4-1). Such nonlinear regression equations are shown in Figure 4-2, where the exponent of the nominal diameter results in -0.26; differing from the value of -0.40 found by Werner and Siebert (1991) (see section 4.1). The slightly lower nail strength in comparison to the wire strength is caused by its calculation with the nominal diameter and not with the (smaller) inner diameter d_i.

51 Again, it can be discussed how the tensile strength of the nail should be calculated. Here, the nominal diameter is used as in practice, only this diameter is known. Moreover, tests have shown that non-smooth shank nails do not always fail exclusively where the nail diameter is smallest, i.e. at d_i, see Figure 4-22. This holds in particular for stainless steel nails.

52 Differently to self-tapping screws, nails made of carbon steel are usually not hardened.

The crosses in Figure 4-2 demonstrate the significant difference between wire strength and subsequent tensile strength of hot-dip galvanised (hdg) nails, especially for smaller diameters where a major part of the cross-section is affected by heat[53]. Especially for hot-dip galvanised nails, wire strength is not correlated to the tensile strength of the finished nails, see also Figure 4-3 on the right. The same applies to hardened nails (special nails), which are hardened only after forming. Because of the rather weak general correlation between wire and nail strength (R = 0.65) shown in Figure 4-3 on the right, it can be stated that wire tests are obsolete. Tensile tests on nails are sufficient to guarantee tensile properties where the strength can be calculated with the nominal diameter. Producers may need wire tests however in order to check delivered steel grades. Finally, it must be emphasised that all comparisons here include material from probably different batches, because tests were carried out on wires and nails produced from, most certainly, different wires. This explains also why mean values are used as individual wire tests cannot be linked to individual nail tests. Furthermore, tensile tests are carried out on specific nail types of one diameter, but only a limited amount of different nail lengths is used. This means that experimentally derived tensile capacities are valid also for nails with different lengths than those tested, which, as discussed in section 4.4, may lead to higher scatter, see Figure 4-5 on the right.

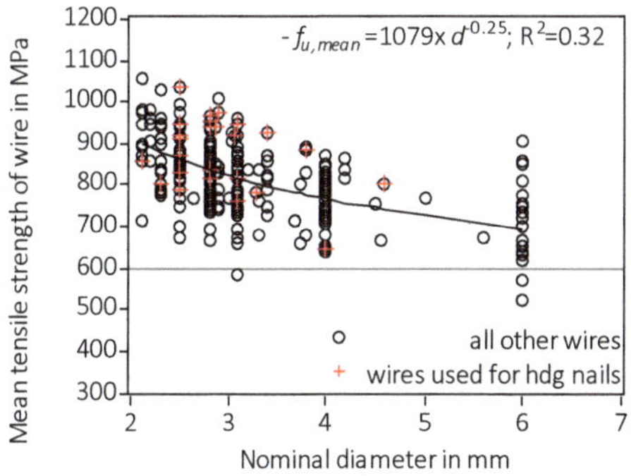

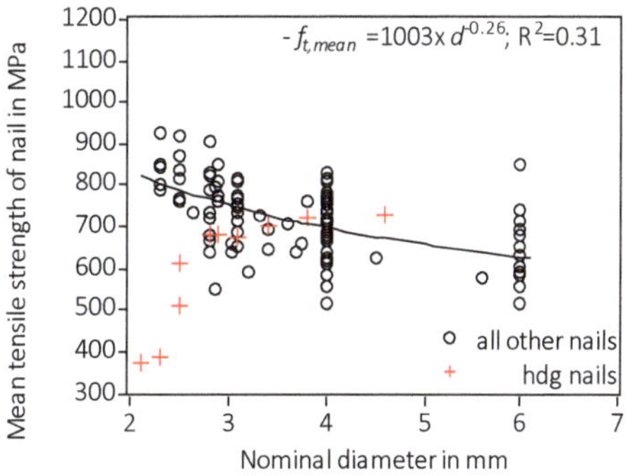

Figure 4-2 Left: Mean tensile strength of wire f_u versus nominal nail diameter d, regression includes 257 test series with 1076 individual values f_u. Right: Mean tensile strength of nail $f_{t,mean}$ calculated with F_t and nominal diameter d, regression excludes hot-dip galvanised (hdg) nails and is based on 104 test series with 982 individual values F_t.

53 Indeed, this effect of lower strength of hdg nails is less strong if Eq. (3-1) is used to derive a "yield moment stress σ_{My}".

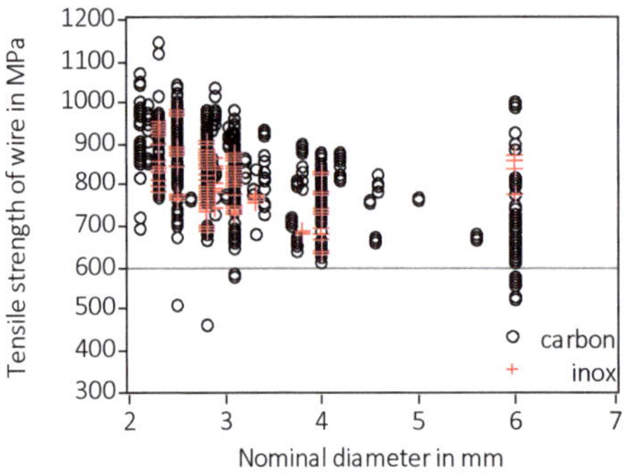

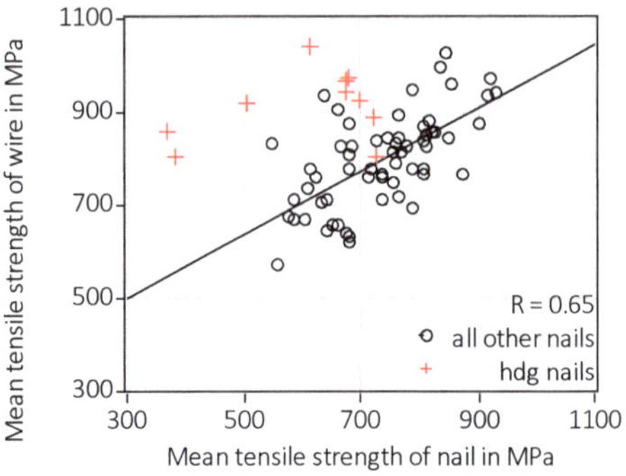

Figure 4-3 Left: Tensile strength of wire (1076 tests) of carbon and stainless steel nails. Line indicating a tensile strength of 600 MPa is shown. Right: Mean tensile strength of wire in dependence of mean tensile strength of nail. Wires that were used to produce hdg nails are identified. Shown trendline with R = 0.65 excludes hdg nails.

4.3.2 Characteristic tensile capacity

The discussed issues concerning variations due to production techniques, different batches and nail lengths influence also the evaluation of characteristic tensile capacities, which are needed for design. The tensile capacity of a certain nail type with non-smooth shank is usually established via testing; it cannot be done otherwise due to e.g. the above-mentioned influence of production. And always, only some nails, typically ten, will be tested and the results will be extrapolated to all other nails of the same type. Therefore, characteristic values need to be calculated taking a certain coefficient of variation and the number of available test values into account. EN 14358 (2016) regulates this. There are two ways to determine the characteristic values to be used in design. Firstly, nail producers declare characteristic values for their nails and designers will need to look those up in the DoP. Secondly, "characteristic nail strength classes" could be defined, with the nail strength depending on the nominal diameter to reflect the increase in tensile strength with decreasing diameter. Designers could then use the values of the relevant class without any need to consult producers' documents. Most probably, a combination of both is best; designers use characteristic strength values defined in classes and at the same time, producers are still free to produce nails with a verifiable higher characteristic tensile strength, e.g. hardened nails. The aim to define strength and not capacity classes is due to the need to provide diameter-independent values as otherwise, each possible nail diameter would need a proper characteristic capacity. Furthermore, as will be seen in section 4.4, strength values may be needed for further calculations.

Figure 4-4 shows the mean tensile strength on the left and the characteristic tensile strength on the right. The empirical standard deviations shown in Figure 4-5 on the right were used to calculate the characteristic values. It can be seen that the difference between mean and characteristic values is not considerable, except for some individual values. This does not come as a surprise seeing that many empirical standard deviations are rather small. The larger empirical standard deviations observed for nails with a diameter of 6 mm are instead reflected through a considerably larger scatter of the characteristic values, where some series see a significant drop between mean and characteristic value. As discussed also in section 4.4.1, maybe a larger threshold value for the standard deviations should be prescribed than the one currently given in EN 14358, i.e. larger than 0.05.

To conclude, a possible classification of characteristic tensile strength values for nails could look as shown in Table 4-2.

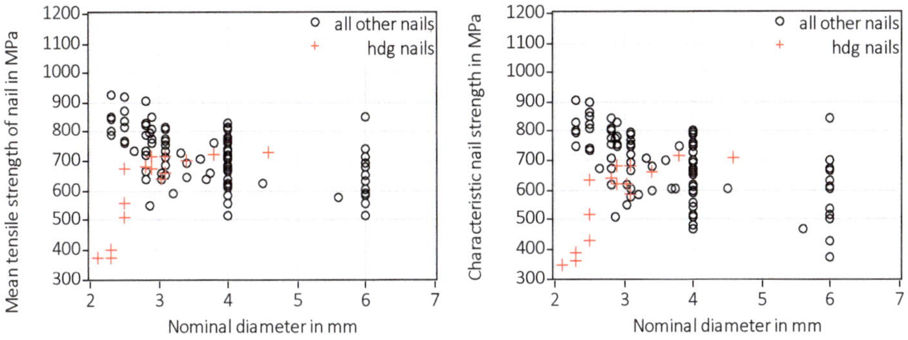

Figure 4-4 Left: Mean tensile strength of nail $f_{t,mean}$ calculated with F_t and nominal diameter. Right: Characteristic tensile strength, calculated acc. to EN 14358 (2016) assuming a logarithmic distribution and the empirical standard deviations. The number of tested values was considered via the k_s-factor (see EN 14358).

Table 4-2 Examples for possible classes of characteristic tensile strength values for nails

Class	I	II	III	IV
Characteristic tensile strength in MPa	300	400	500	700
Examples for nails that belong to this class	hdg nails	Nails with diameter $d \geq 6$ mm	Nails with diameter 2.5 mm $< d <$ 6 mm	Nails with diameter $d \leq 2.5$ mm

4.4 Yield moment

4.4.1 Nonlinear regression

Introduction

The yield moment and the tensile strength of the nail are the two steel properties needed for joint design in accordance with the current Eurocode 5 (EN 1995 1-1, 2010). Yield moment and tensile strength are interdependent; a high tensile strength will lead to a high yield moment. The yield moment obviously depends on the diameter; the larger the diameter, the larger is the yield moment, see Figure 4-5 on the left. The yield moment can be made independent of the diameter by calculating a "yield moment stress σ_{My}" of the nails where σ_{My} is calculated in accordance with Eq. (3-1). Figure 4-5 on the left also shows the strong influence of a hardening procedure, whereas no significant difference can be seen between stainless steel and carbon steel nails. To analyse the relationship between the yield moment M_y and the tensile capacity F_t of the nail, a regression analysis can be carried out. As M_y and F_t are determined using different nails and sometimes even nails from different batches[54] (e.g. nails of the same diameter but of different length were tested), a regression analysis based on individual dependent and independent variables is not possible. Therefore, test results for M_y and F_t have to be compared as mean and not as individual values.

When performing a regression analysis, the quality of the underlying database must be thoroughly examined and outliers with studentised residuals larger than |3| should be excluded from further analysis (Hartung, 2009, p. 586). As a first step, the standard deviations of the individual test series are investigated. Figure 4-5 on the right shows the standard deviations of all test series for M_y and F_t. Mean standard deviation for both parameters is 0.03. Concerning scatter, M_y and F_t do not differ significantly. 14% (F_t) and 15% (M_y) of the tests had standard deviations larger than 0.05, which is the threshold value given in EN 14358 (2016). These high values for the standard deviation may be due to the determination of M_y and F_t on nails from different batches. For instance, nails of the same diameter but with a different length were used within single series. In practice, characteristic values per nail type and diameter will always be determined using a limited amount of different lengths, leading to a small standard deviation. Usually one length is considered being representative for all lengths of the same nail type and diameter. The evaluated characteristic values will then be extrapolated to all available lengths of the same nail type and diameter. Consequently, again the question arises if the minimum value for the standard deviation of 0.05 given in

54 This general issue of the database has many implications which will be discussed in the following, and which led to the small experimental campaign discussed in section 4.4.4.

EN 14358 (when using a lognormal distribution) is sufficiently reliable or if for instance, higher values are required to represent data from different batches.

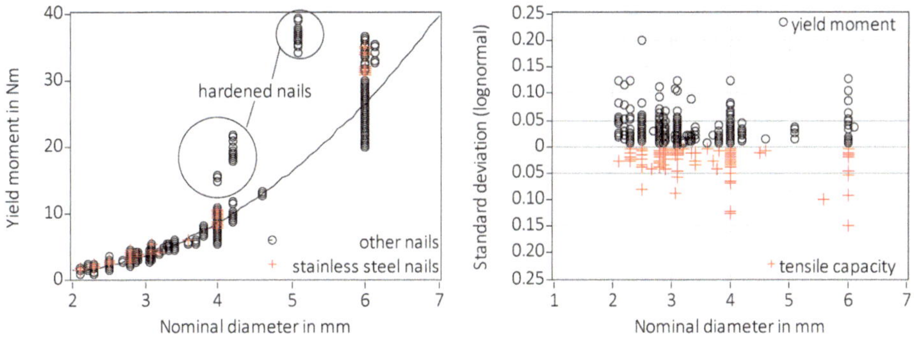

Figure 4-5 Left: Yield moment (2854 tests) of carbon and stainless steel nails. Hardened nails are identified. Right: Standard deviation s_y of 353 test series for M_y and 119 test series for F_t.

As already stated, the determination of M_y and F_t using different nails impedes a direct correlation between individual test values. However, both M_y and F_t are steel properties and there are no differences in the variation except for those related to the test setup or random variations. In order to increase the size of the dataset, one could argue that mean tensile capacities $F_{t,mean}$ could be compared to individual yield moments M_y. This in turn would mean that multiple dependent variables M_y are linked to one mean independent, or, better term in this context, explanatory variable $F_{t,mean}$, which leads to uncontrollable effects in a regression analysis as illustrated in Figure 4-6. Expected values resulting from a nonlinear regression analysis (similar to Eq. (4-4)) using individual M_y-values and a mean tensile capacity[55] per test series correlate well with test results, Figure 4-6 on the left. If showing studentised residuals versus experimental values however, Figure 4-6 on the right, the effect of explaining multiple M_y-values with one $F_{t,mean}$-value can be seen. Per explanatory variable $F_{t,mean}$, one expected value $M_{y,exp}$ is established. The residuals are tilted to the right as within one test series, multiple individual M_y (which are different in size) are compared with one $M_{y,exp}$ to obtain residuals.

55 Obviously, the regression is performed with the mean tensile strength $f_{t,mean}$, calculated in accordance with Eq. (4-3), using $F_{t,mean}$ and d_{meas}.

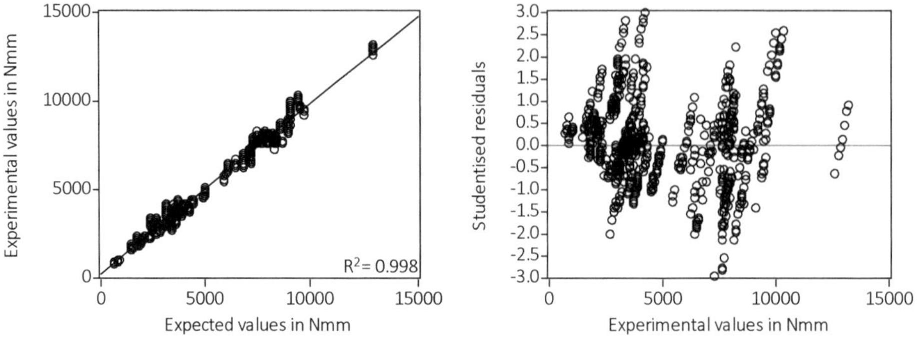

Figure 4-6 Result of a nonlinear regression based on individual values of M_y and mean values of F_t. Outliers were excluded and only data up to 15 Nm is shown. Left: Expected versus experimental yield moments. Right: Experimental M_y-values versus studentised residuals.

The variables on both sides of the regression equation should hence have a unique relationship; i.e. a certain value of the dependent variable should be explained by one value of the independent variable(s). From this, another question arises, i.e. which nail diameter should be used in the regression analysis, the "measured nominal diameter d_{meas}" or the "declared nominal diameter d". To obtain a more significant regression, d_{meas} is used as d_{meas} is the real diameter of the nail that was tested. When using the "declared nominal diameter d", errors due to differences between declared and measured diameters are included in the regression residuals. Finally, the regression analysis, whatever shape it takes, is deterministic as test results are considered as both dependent and independent variables.

Nonlinear regression

As a first step in this analysis, the tensile capacity F_t is transformed into a tensile strength f_t, using the ***measured*** nominal diameter d_{meas}, Eq. (4-3). Such a procedure is expedient as the ***declared*** nominal diameter (and e.g. not the inner diameter d_i) is used for design purposes. Moreover, other diameters, e.g. d_i, than the nominal are usually unknown respectively were not recorded in the database.

$$f_t = \frac{4 \cdot F_t}{\pi \cdot d_{meas}^2} \qquad (4\text{-}3)$$

where

f_t	Tensile strength of nail in MPa
F_t	Tensile capacity of nail in N, test value
d_{meas}	Measured nominal diameter in mm

The subsequent regression equation takes the following shape:

$$M_{y,mean} = \alpha \cdot f_{t,mean} \cdot d_{meas}^{\beta} \tag{4-4}$$

where

$M_{y,mean}$	Mean yield moment in Nmm, dependent variable
$f_{t,mean}$	Mean tensile strength of nail in MPa, independent variable
d_{meas}	Measured nominal diameter in mm, independent variable
α, β	Regression parameters

The shape of Eq. (4-4) is not randomly chosen. The equation to calculate the full plastic bending capacity of a circular cross-section has a similar shape, see Eq. (3-2). Moreover, all current design equations for different fasteners have a similar shape, see e.g. Eq. (3-4) for staples and Eq. (4-1) for smooth shank nails.

Different database compositions were investigated in order to judge the quality of the overall regression and to answer the question, which test series should be included in the analysis. Regression analyses on subsets[56], e.g. considering only certain nail types or stainless steel nails, resulted in differences of less than 3% of the expected value for M_y compared to the overall regression given in Eq. (4-5) that included all subsets. Therefore, no differentiation was made within the database[57], e.g. with respect to different nail types or carbon and stainless steel. The hardened nails highlighted in Figure 4-5 on the left were not included in the analysis, since only M_y was tested and no data was available to calculate f_t.

The residuals were generally rather heavy-tailed and all outliers with a studentised residual larger than |3| were excluded (Hartung, 2009, p. 586). In total, 6 test series were excluded. Finally, the regression was based on 99 test series (with 979 M_y- and 986 F_t-values) and resulted to:

$$M_{y,mean} = 0.1772 \cdot f_{t,mean} \cdot d_{meas}^{3.0191} \tag{4-5}$$

where

$M_{y,mean}$	Expected value of mean yield moment in Nmm
$f_{t,mean}$	Mean tensile strength of nail in MPa, independent variable
d_{meas}	Mean measured nominal diameter in mm, independent variable
e	Residuals

56 Not all subsets were statistically relevant; e.g. only 6 series for spiral nails were available.

57 Furthermore, analyses on subsets do not deliver significant results as most subsets contain only few individual test results (e.g. only 15 yield moments for galvanised spiral nails with a diameter of 4.2 mm).

Table 4-3 gives the key data for some regression analyses. The first two columns give data for analyses in accordance with Eq. (4-4), but considering declared nominal diameters d, also when calculating the tensile strength f_t. The last two columns instead represent regressions where measured nominal diameters d_{meas} were used. In both studies, results including and excluding outliers are given. The effect of the outliers can be immediately seen when considering the reduced standard deviation *std* although the (pseudo-)coefficients of determination R^2 did not change much. It can be furthermore seen that using the measured diameters adds additional scatter in comparison to the regression analyses using declared diameters. Most importantly, the magnitude of the regression parameters did not change significantly and lead to similar expected values with differences of only ca. 3%. This finding is valid also for other regression analyses, e.g. considering characteristic values (see Sandhaas and Görlacher, 2018) or the above discussed regression based on individual M_y- and mean F_t-values, and it indicates that, in general, the regression delivers reliable results, with regression parameters that are very similar to the numeric values of Eq. (3-2).

Table 4-3 Regression analysis with and without outliers. *std* = standard deviation of residuals in Nmm. In parenthesis are the 95% confidence limits of the regression parameters α and β.

	Using declared nominal diameter d		Using measured nominal diameter d_{meas}	
	Including outliers	Excluding outliers	Including outliers	Excluding outliers
n	105 (1034 M_y, 1035 F_t)	100 (979 M_y, 970 F_t)	105 (1034 M_y, 1035 F_t)	99 (979 M_y, 986 F_t)
α	0.185 (0.171-0.198)	0.177 (0.167-0.186)	0.187 (0.172-0.201)	0.177 (0.166-0.188)
β	2.992 (2.948-3.036)	3.022 (2.989-3.055)	2.984 (2.937-3.030)	3.019 (2.982-3.056)
R^2	0.995	0.997	0.995	0.996
std	499	361	535	398

Residual analysis

Figure 4-7 shows the experimental versus the predicted values (Eq. (4-5)) on the left, and the qq-plot of the studentised residuals on the right[58]. It can be seen that the residuals are approximately normally distributed and therefore, the general assumption of any regression analysis based on the method of least squares holds. Looking at Figure 4-8 on the left, residuals seem to be homoscedastic as required, i.e. the residuals have the same variance. As homoscedasticity is difficult to detect due to the small sample size, parameters could be estimated by weighted nonlinear least squares alternatively. Standard statistical tests available in SAS to check the validity of the regression hold, e.g. the F-value test for the regression or goodness-of-fit tests for normal distribution of residuals. Also skewness is close to zero

58 All discussions are based on analyses using the measured nominal diameter d_{meas} and excluding outliers.

and bias is less than 1%, which means that the regression parameters have a close to linear behaviour, and calculated standard errors and confidence intervals can be used. If looking further into residuals however, first drawbacks arise that cannot be detected without thoroughly examining the residuals. Standard methods available in statistics packages such as SAS do not easily reveal these issues. In other words, standard statistical checks should not be trusted blindly.

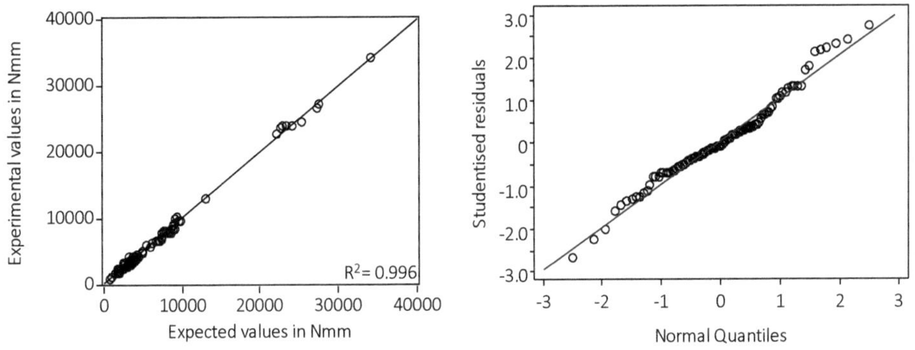

Figure 4-7 Left: Experimental and predicted values (Eq. (4-5)) for the yield moment $M_{y,mean}$. Right: qq-plot of studentised residuals.

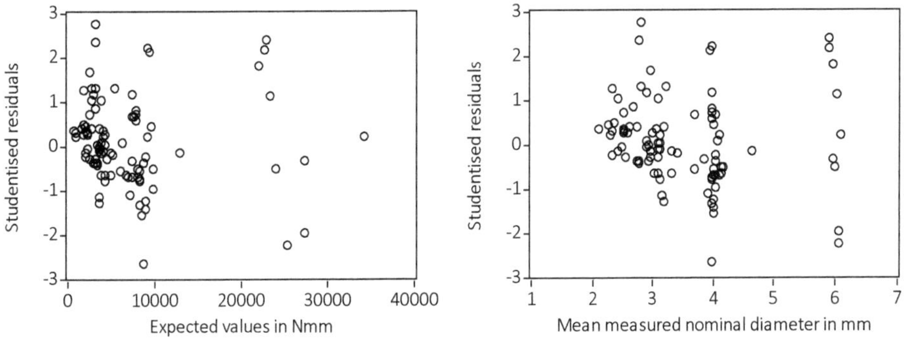

Figure 4-8 Studentised residuals versus expected value (left) and versus measured nominal diameter (right).

Figure 4-8 on the right shows the measured nominal diameter d_{meas} versus the studentised residuals and it can be seen that the scatter of residuals is smaller for small diameter nails. This uneven distribution of the residuals cannot be easily detected in e.g. Figure 4-8 on the left. Different hypotheses concerning the non-constant variance of the residuals are possible. First of all, it may mean that indeed residuals are not evenly distributed and that the non-constant variance must be considered in further analyses. The findings of section 4.3.1 could be the reason for this non-constant variance, i.e. that the tensile strength of the nails is decreasing with increasing diameter. This relationship between strength and diameter is however not reflected in e.g. the variance inflation factor (VIF) of f_t and d_{meas} that amounts to 1.06. The VIF is a measure of collinearity and a VIF of below 10 is commonly considered as indicating that there is no collinearity (Chatterjee and Price, 1995, p.202[59]). This confirms yet again the previous statement that the outcomes of standard statistical procedures should not be the sole validation tools, but that a thorough analysis of residuals using graphical representations and mechanical knowledge is a prerequisite to obtain valid final results.

A second hypothesis is that the database is not representative enough. Indeed, 11 test series were available for e.g. 2.1 mm nails and 30 for 6 mm nails, and these numbers indicate that not enough test results are available for small diameter nails. From this follows that, if sufficient test series for all diameters would have been available, the studentised residuals for all diameters would oscillate around ± 3. Without further evidence, however, this remains a hypothesis.

Discussion on outliers

The discussion up to now considered only the regression analysis where outliers with studentised residuals of |3| were excluded. The question remains if it is correct to exclude these. The analysis of outliers shows that many outliers are due to the insufficient quality of the available datasets with resulting large scatter. It is usually not clear if these values result from e.g. measuring errors and should be excluded or if they are correctly determined, represent the scatter of the respective property and should hence not be excluded. Table 4-4 illustrates this. It gives individual values for both F_t and M_y of a test series that was excluded. The individual value in bold is responsible for the deletion; if this value of 12.50 Nm is deleted, the studentised residual drops below |3|. Looking at the other values for M_y in Table 4-4, above all the value of 12.70 Nm, it seems that the value of 12.50 Nm was properly determined and should hence not be excluded. If looking further, it can be seen that 60 mm long ringed shank nails were used to determine M_y and 100 mm long ringed shank nails for F_t. This means that the independent and dependent variables of Eq. (4-5) were evaluated on different nails. As the regression aims at establishing a relationship between M_y and F_t of one nail, it is correct to delete both series as they are not interdependent. In other words, not excluding them would mean that the regression links two series, which in reality are not linked. This general issue of the database can also be seen in Figure 4-5 on the right that shows the sometimes rather high standard deviations, which often affect series where nails with different lengths were tested.

59 Uncorrelated independent variables are also a prerequisite of a regression analysis.

Table 4-4 Example of an excluded test series.

6 mm ringed shank nails, F_t in kN, L = 100 mm	6 mm ringed shank nails, M_y in Nm, L = 60 mm
21.20	16.00
21.80	12.70
22.20	15.40
22.20	**12.50**
22.60	16.60

However, deleting all series where properties were determined on different nails is no solution; for two reasons. First, only very few test results would remain in the database in that case, also because all doubtful series should be deleted as well where the length is not given and where hence it is not clear if the "same" nails were tested. The second reason has been discussed before and has practical implications. Usually one length is considered being representative for all lengths of the same nail type and diameter and also in future, it will be impossible to test all available nails. Furthermore, already a new batch during production may lead to different results. The required testing effort would be immense and hence the need of testing nail groups instead of all nail types (and all batches) will remain. Finally, both properties are always determined using different nails; the tensile capacity on one nail and the yield moment on another. Therefore, data must represent nails from different batches. Here, Eq. (4-5), excluding outliers, remains nevertheless the reference regression result.

Additionally, difficulties concerning the correct experimental determination of the yield moment M_y as outlined in Steilner and Blaß (2014) must be discussed here as well (see also section 5.3.2). Due to these difficulties, the observed scatter for M_y (see Figure 4-5 on the right) is expected to be higher than it is in reality; above all, if experimental results from different testing bodies are included. There is no means to assess this additional scatter which is solely due to the test setup and measuring procedure.

Conclusion

The observed issues around the available database stem from the fact that only retrospective data is available and that properties are always determined on different nails. All collected data stem from certification tests, which did not follow a common, defined purpose. In a strict mathematical way, the prerequisites of correct regression analysis are not fulfilled, i.e. that M_y and F_t are determined using the same nail. This general issue could be resolved by simply carrying out innumerable additional tests. In the following sections however, the aim is rather to investigate how the evaluated database can be used to

- generate representative databases including realistic distributions (section 4.4.2),
- reliably determine characteristic values (section 4.4.3).

For this purpose, the stochastic approach used in Frese et al.(2010) to simulate withdrawal capacities is applied[60]. In all following analyses (sections 4.4.2 and 4.4.3), the database excluding the identified outliers was used. Finally, additional test series were carried out, where both yield moment and tensile strength were evaluated using one single nail (section 4.4.4). The results of these additional tests are discussed with a strong focus on the implications around the original retrospective database.

4.4.2 Generation of empirically represented data

Introduction

The basic idea is to generate empirically represented data to boost experimental datasets and to finally derive characteristic values resp. to generate databases for further reliability analyses (e.g. similar to Jockwer, 2016). If dependent and independent parameters of a regression analysis have a unique relationship (i.e. the parameters stem from one single test[61], or, at least, from tests on the same nails), then only data representing the residuals *e* must be simulated – in our case, the residuals *e* of Eq. (4-5). A prerequisite to this is that the test results used in the regression analysis are representative. For instance, if withdrawal capacities are determined using always the same piece of timber, obviously any regression analysis is meaningless as the density does not vary.

There is scope to assume that the test results used here are indeed not representative. In Eq. (4-5), M_y is a function of the tensile strength f_t and the nail diameter *d*. The scatter of f_t is not fully represented as not enough f_t-values were determined using different nail lengths, and therefore the residuals *e* do not contain enough information about the scatter of f_t. Table 4-5 illustrates this, where individual values of one series are given. The standard deviation of the series shown in Table 4-5 is 0.07 and tensile capacities determined using 100 mm nails are higher than those determined using 60 mm nails. Generally, the length of the used nails is unknown and it may well be that those series with large standard deviations (Figure 4-5 on the right) were carried out using various nail lengths. Furthermore, if f_t-values are simulated, it must be considered that low diameter nails can reach higher tensile strengths f_t, see Figure 4-2 on the right. Recalling Figure 4-8 on the right, also studentised residuals do not seem to be independent of the diameter. As discussed previously, various assumptions could hold here. One is that the dependence of f_t on the diameter influences the distribution of the residuals or, more probable, not enough data is available, i.e. the database is, again, not representative.

60 Simulating data is not new and was applied e.g. to generate MOE data for timber (Taylor and Bender, 1989).

61 This is possible e.g. for withdrawal tests. Here, results from two different tests are considered, M_y and F_t.

Concerning the diameter d, maximum differences between measured and declared nominal values (expressed as Δd = ratio of measured over declared nominal diameter) amounted to a maximum of 14%, see Figure 4-9 (96% of the series had deviations of less than ±5%). As measured nominal diameters were used in the regression analysis, these deviations are included in the residuals.

Table 4-5 Tensile capacity F_t in kN on 4 mm ringed shank nails, where nails with different lengths were used.

L = 60 mm	L = 100 mm
8.51	10.50
9.23	9.64
9.64	10.50
9.42	10.20
9.54	10.50

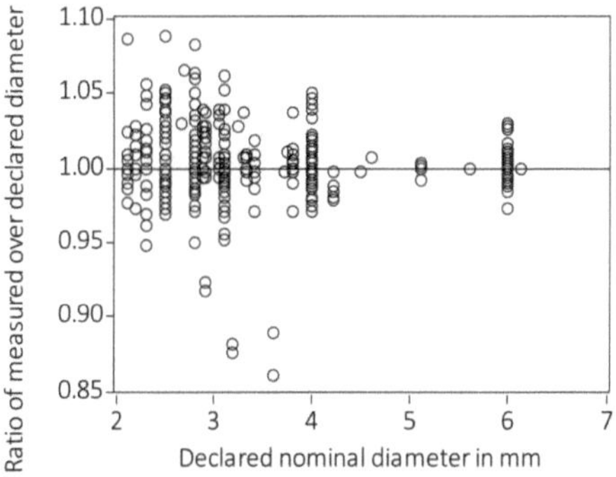

Figure 4-9 Δd-values versus declared nominal diameters.

Using simulated data and Eq. (4-5), "representative" M_y-values can be calculated that, due to the use of simulated data, better represent the observed variability due to e.g. different nail lengths. Direct generation of empirically represented M_y-data, however, is only possible per diameter. This would lead to unreliable data as e.g. for a diameter of 3.3 mm, only 15 individual M_y-values are available, and no M_y-values are available for other than the tested nail diameters. Within the existing database, only for 2.1, 2.3, 2.5, 2.8, 2.9, 3.1, 4 and 6 mm nails, more than 50 individual results are available. Generation of empirically represented M_y-values is hence not possible without further large experimental campaigns.

As Eq. (4-5) showed that M_y in principle is a function of f_t and d and a general equation covering all nail types and diameters is possible, simulated data representing residuals and tensile strength (**Simulations A**, **B** and **C**) are used in further analyses:

- **Simulation A**: An even oscillation of studentised residuals between ± 3 is assumed, postulating that all residuals would oscillate between ± 3 if a comprehensive database would have been available.
- **Simulation B**: The observed dependence of the studentised residuals on the diameter (Figure 4-8 on the left) is considered.
- **Simulation C**: Apart from simulated values of the residuals *e* generated in **Simulations A** and **B**, additionally, f_t-values are simulated that depend on the nail diameter (Figure 4-2 on the right).

Generation of empirically represented residuals

Simulation A (constant variance of studentised residuals)

Using data simulation techniques[62], the residuals *e* of Eq. (4-5) and shown in Figure 4-10 on the left are modelled. As can be seen in Figure 4-10 on the left (and also in Figure 4-7 on the right), these residuals *e* are approximately normally distributed where however only 99 values are available. Figure 4-10 on the right shows simulated residuals assuming a normal distribution (*mean* = 30.47, *std* = 398.41). It is noteworthy to mention that the simulated minimum and maximum values are significantly larger than the values from the regression analysis whereas the simulated 1st and 99th percentiles correspond rather well to the minimum and maximum residuals from the regression.

Simulation B (non-constant variance of studentised residuals)

If the non-constant variance of the residuals observed in Figure 4-8 on the right is assumed to be representative of the truth, the simulation approach remains methodically the same, but multiple simulations must be run by grouping residuals of the regression analysis in dependence of the diameter. Then, per group, values are simulated using the respective mean values and standard deviations. For illustration purposes and re-considering Figure 4-8 on the right, two groups are formed, and the separator is a mean measured diameter of 3.9 mm[63]. To simulate residuals for nails with diameters $d < 3.9$ mm, the normal

[62] The ***streaminit*** function (univariate data) and ***simnormal*** procedure (multivariate data) of SAS 9.4 are used here.

[63] The two "outliers" at around a diameter of 3 mm and with standardised residuals larger than +2 are eliminated (Figure 4-8 on the right). This is done to show the effect of smaller residuals for smaller diameter nails in further analyses.

distribution shown in Figure 4-11 on the left is used. Residuals for nails with $d \geq 3.9$ mm are simulated using the normal distribution in Figure 4-11 on the right. The difference in variance for small and large diameter nails can be clearly seen in Figure 4-11, having the same horizontal scale. The results of the simulations are given in Figure 4-12.

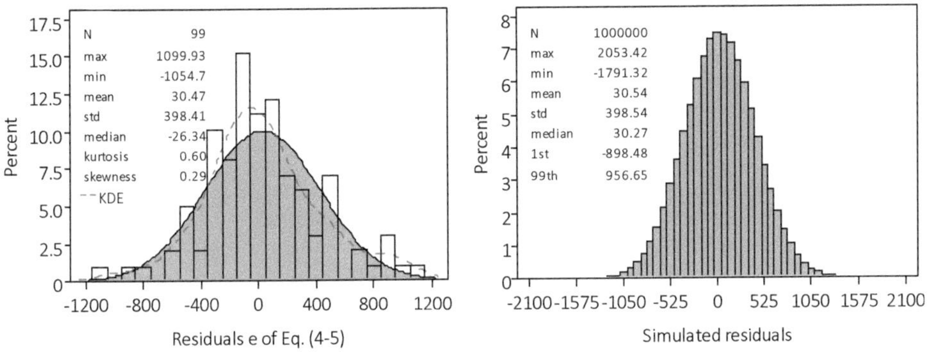

Figure 4-10 Left: Histogram of residuals *e* of Eq. (4-5) with fitted normal distribution. Right: Histogram of simulated residuals; using mean and standard deviation from the histogram on the left.

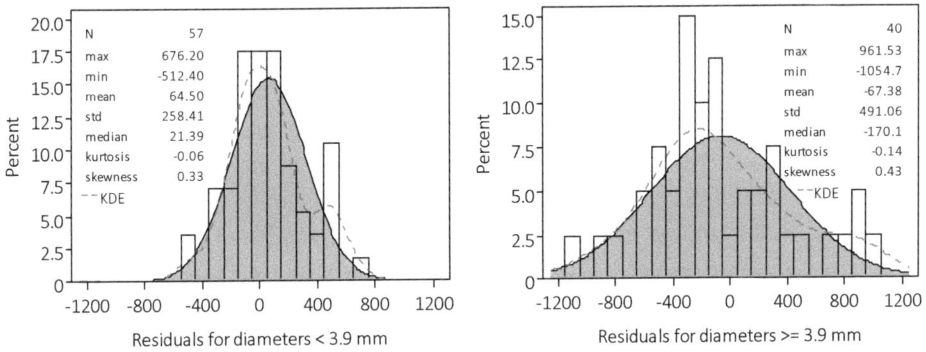

Figure 4-11 Left: Histogram of residuals *e* of Eq. (4-5) with fitted normal distribution and for nails with $d < 3.9$ mm. Right: Histogram of residuals *e* of Eq. (4-5) with fitted normal distribution and for nails with $d \geq 3.9$ mm.

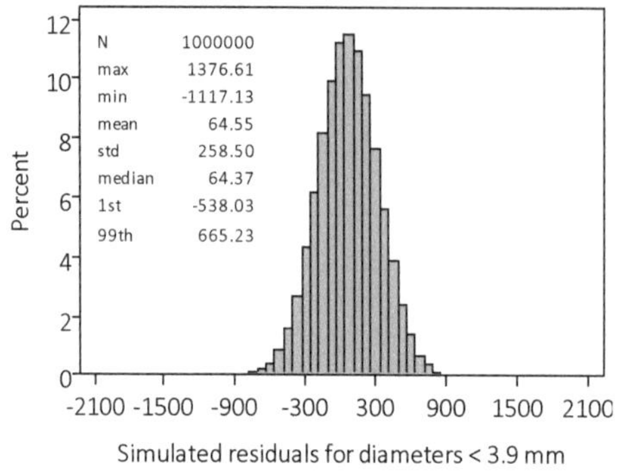

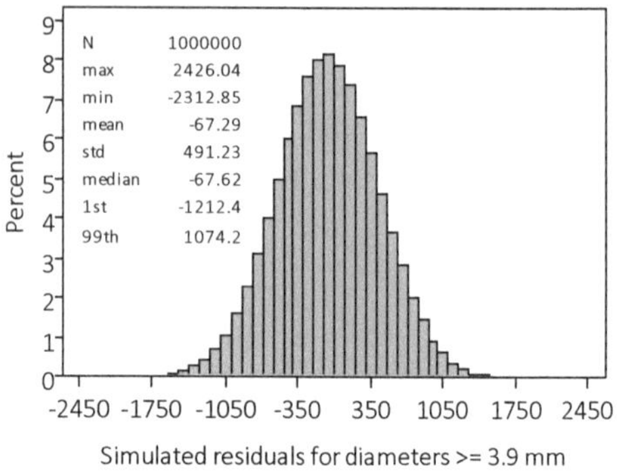

Figure 4-12 Left: Histogram of simulated residuals for nails with d < 3.9 mm. Right: Histogram of simulated residuals for nails with d ≥ 3.9 mm (N.B. different scale of horizontal axis).

Generation of empirically represented tensile strength

Simulation C

Generating data only for residuals for later calculation of yield moments (section 4.4.3) neglects the fact that also the tensile strength f_t scatters, and hence, that also for f_t, empirically represented data should be generated. As f_t depends on the nail diameter, multivariate data must be simulated. Multivariate simulation tools based on normal distributions are given in SAS and are applied here (***simnormal*** procedure). Pearson's correlation coefficient for *d* and f_t[64] is R = -0.54, and the normal distribution parameters for f_t are *mean* = 729.5 MPa and *std* = 102.4 MPa. Hot-dip galvanised nails were excluded. Figure 4-13 shows the simulation results $f_{t,sim}$. On the left, the correlation between tensile strength and diameter can be seen where the simulated values are grouped by diameters and all simulated values with d < 1.45 mm were deleted. Analogously to the simulated residuals, the simulated minimum and maximum tensile strength values are larger than test values (where min = 355 MPa, max = 1004 MPa).

In the following, the two exemplary diameters *d* = 2.8 mm and *d* = 6 mm are chosen to have a closer look on the results[65]. Simulated and experimental tensile strength values for a nail diameter of 6 mm are shown in Figure 4-14, and Table 4-6 gives statistical data for both exemplary nail diameters. The number of simulated values per diameter is smaller here as the total 1 million simulated $f_{t,sim}$-values are distributed to the different nail diameters as shown in Figure 4-13 on the left, and hence, per diameter, less than 1 million values are available. Simulated 5th percentiles are lower in comparison to observed experimental val-

64 f_t was calculated with the measured nominal diameter d_{meas}.

65 In the database, the most frequent diameters were 2.5, 2.8, 3.1, 4 and 6 mm, see section 4.2.

ues. The characteristic values in accordance with EN 14358 (2016) have a considerable scatter (see also Figure 4-4 on the right), and the minimum value is smaller than the 5th percentiles derived from simulated data. Generally, the simulated values fit well to the experimental data, and small experimental databases can be extended using multivariate procedures in order to obtain larger databases for further analyses. Consequently, simulating data are a good tool to close experimental gaps in a swift and easy-to-use manner.

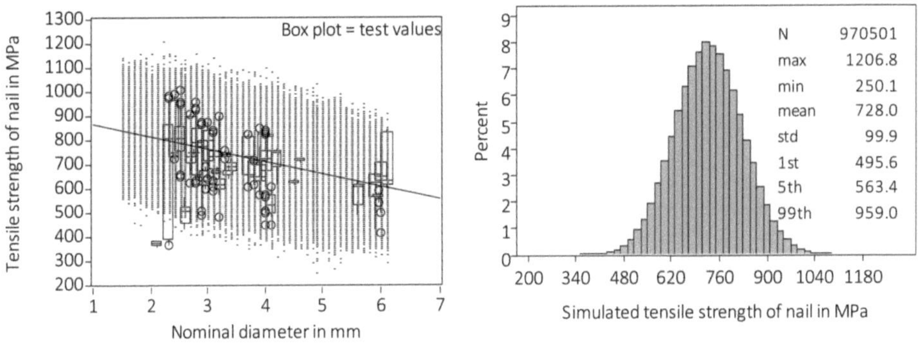

Figure 4-13 Left: Tensile strength in dependence of nominal diameter, box plot shows test values, grey dots show simulated data and their trendline is shown in red. Right: Simulated tensile strength values for nails.[66]

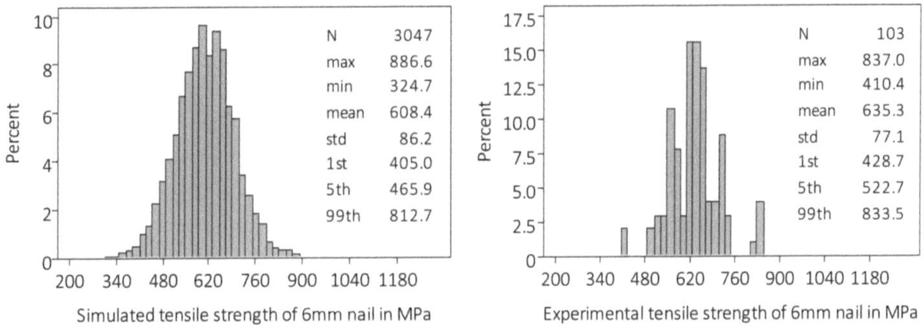

Figure 4-14 Tensile strength values for nails with a diameter of 6 mm. Left: Simulated values. Right: Experimental values.

[66] The generation of multivariate random numbers based on normal distribution led to negative values that were deleted and therefore, n < 1 million.

Table 4-6 Simulated and experimental tensile strength values f_t in MPa.

	2.8 mm nail		6 mm nail	
	Simulation	Test	Simulation	Test
n	28076	100	3047	103
mean	772	762	608	635
min	406	619	325	410
max	1160	955	887	837
1st percentile	570	624	405	429
5th percentile (observed value)	629	633	466	523
99th percentile	970	945	813	834
Characteristic values in accordance with EN 14358 (lognormal distribution, observed *std*, k_s)		616 to 847		370 to 844

4.4.3 Calculation of yield moments based on simulated data

Simulations A, B and C

Simulation A (Constant variance of studentised residuals)

Simulated expected values for the yield moment $M_{y,sim}$ can now be calculated based on Eq. (4-5). As discussed above, different assumptions can be taken to evaluate $M_{y,sim}$. As a first step, it is assumed that the studentised residuals have a constant scatter of ± 3 for all diameters (Figure 4-10) and additional scatter in tensile strength f_t (Figure 4-13) is not considered.

Besides the simulated residuals e, a nominal diameter d and a tensile strength f_t of the nails are needed to calculate $M_{y,sim}$. The results are shown for the two diameters of 2.8 and 6 mm, and f_t is the mean value of the experimentally derived tensile strength per diameter (excluding hot-dip galvanised nails). Inserting all these in Eq. (4-5) leads to the $M_{y,sim}$-values shown in Figure 4-15.

Exemplarily, the approach is outlined in Eq. (4-6) for a nominal diameter of $d = 6$ mm:

$$M_{y,sim} = 0.1772 \cdot 635 \cdot 6^{3.0191} \quad (4\text{-}6)$$

where

$M_{y,sim}$ Simulated value of yield moment in Nmm, for $d = 6$ mm

635 Mean tensile strength of 6 mm nails in MPa, from experimental data and calculated with measured nominal diameter, see Table 4-6

6 Nominal diameter 6 mm

e 1 million simulated residuals given in Figure 4-10

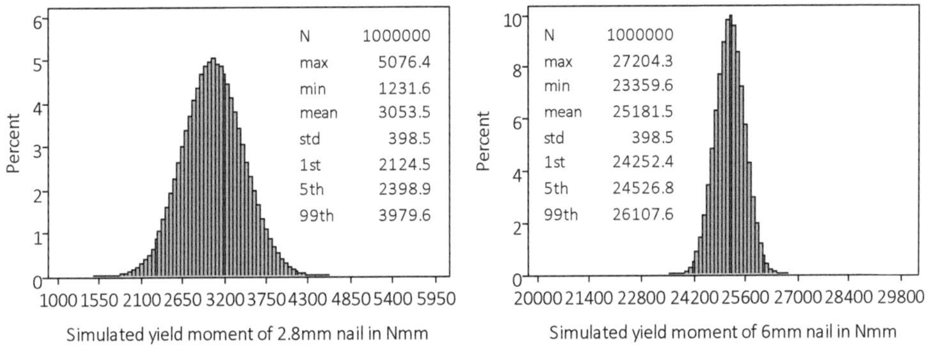

Figure 4-15 Histogram of simulated yield moments considering a constant scatter of ±3 of the studentised residuals. Left: Diameter 2.8 mm. Right: Diameter 6 mm.

Simulation B (Non-constant variance of studentised residuals)

As a next step, the observed dependence of the studentised residuals on the diameter was considered by using the simulated residuals given in Figure 4-12; hence by dividing into two groups with $d < 3.9$ mm and $3.9 \text{ mm} \leq d$. The results are shown in Figure 4-16. Unsurprisingly, the scatter for 2.8 mm nails increases and decreases for 6 mm nails in comparison to Figure 4-15.

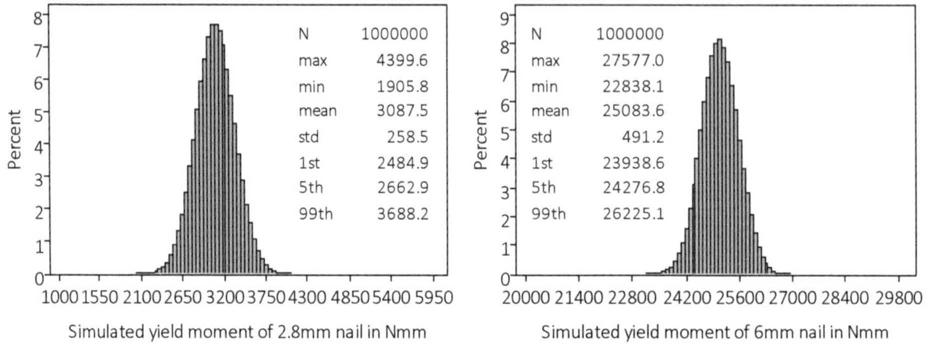

Figure 4-16 Histogram of simulated yield moments considering two different distributions of the simulated residuals (Figure 4-12). Left: Diameter 2.8 mm. Right: Diameter 6 mm.

Simulation C (Simulated f_t-values and simulated residuals e with constant variance)

Finally, also an additional scatter of the tensile strength of nails is considered, using Eq. (4-7) which shows the procedure for 2.8 mm nails, and which will lead to the largest scatter of simulated yield moments. The used simulated tensile strength values $f_{t,sim}$ are shown in Figure 4-13. The used simulated residuals were generated based on an assumed constant variance of ± 3 of the studentised residuals (**Simulation A**, Figure 4-10). The number of simulated values $M_{y,sim}$ per diameter is smaller as fewer $f_{t,sim}$-values are available per diameter, see Table 4-6. The results are shown in Figure 4-17.

$$M_{y,sim} = 0.1772 \cdot f_{t,sim} \cdot 2.8^{3.0191} + e \qquad (4\text{-}7)$$

where

$M_{y,sim}$ Simulated value of yield moment in Nmm, for d = 2.8 mm

$f_{t,sim}$ Simulated tensile strength of 2.8 mm nails in MPa, see Figure 4-13 on the left

2.8 Nominal diameter 2.8 mm

e 1 million simulated residuals given in Figure 4-10

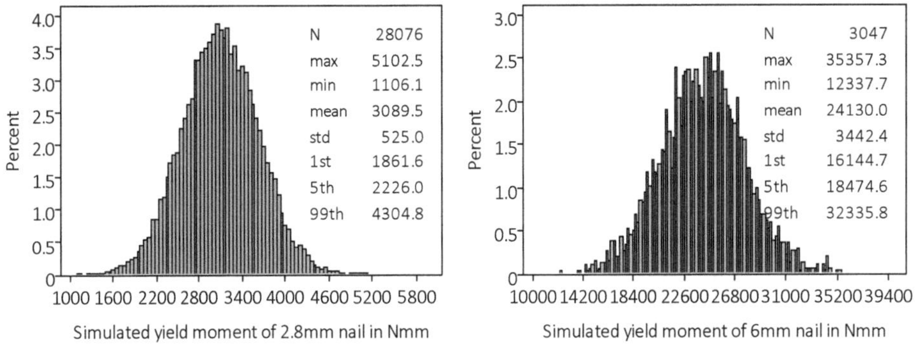

Figure 4-17 Histogram of simulated yield moments a constant scatter of ±3 of the studentised residuals and additional scatter through $f_{t,sim}$. Left: Diameter 2.8 mm. Right: Diameter 6 mm.

Experimental yield moments

Finally, the experimental values for 2.8 and 6 mm nails are shown in Figure 4-18. Interestingly enough, yield moments for 6 mm nails are right-skewed. There is, however, no reason as to why yield moments of 6mm nails should not be normally distributed.

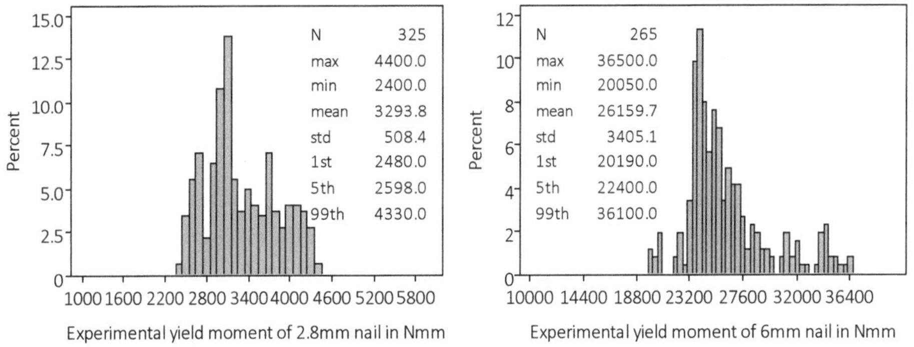

Figure 4-18 Histogram of experimental values of yield moments; hdg nails are excluded. Left: Diameter 2.8 mm. Right: Diameter 6 mm.

Influence of quantity of simulated values

Triggered by the fact that less simulated values were used per diameter when using Eq. (4-7), the influence of the quantity of simulated values on the results was investigated. This is exemplarily shown in Figure 4-19 for a diameter of 4 mm, for which the largest amount of test values was available (n = 630). The effect of using less simulated values can be seen when looking at the extreme values. 1st, 5th and 99th percentiles as well as mean values are very similar however. Hence, about 10000 simulated values are sufficient to properly model yield moments if extreme values are not considered.

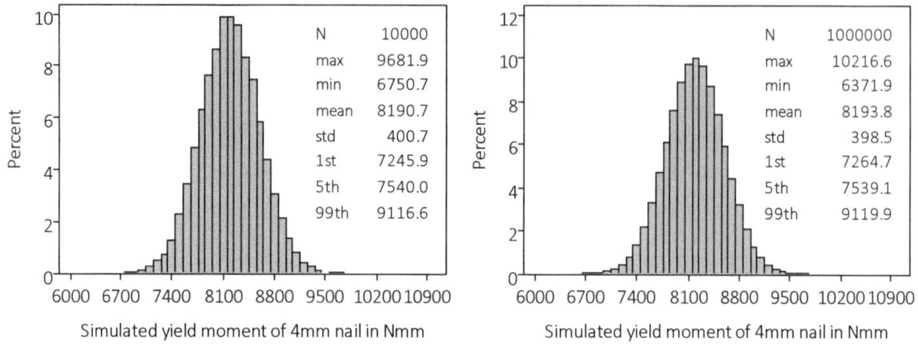

Figure 4-19 Histogram of simulated values of yield moments of 4 mm nails, using a fixed tensile strength and simulated residuals based on a constant scatter of ±3 of the studentised residuals. Left: 10000 simulated values for the residuals were generated. Right: 1 million values were generated.

Comparison of Simulations A, B and C with experimental data

The main statistical data of all simulations (Figure 4-15 to Figure 4-17) and the test results shown in Figure 4-18 are summarised in Table 4-7 for 2.8 mm nails and in Table 4-8 for 6 mm nails. A first observation is that "few" test results are available per diameter; only 325 values for 2.8 mm nails and 265 for 6 mm nails. The scatter of these few results is obvious when looking at the histograms given in Figure 4-18. The simulations seem to be capable of providing a broader set of data (using generated, empirically representative values). When looking at mean values, these tend to get smaller when yield moments are simulated in comparison to experimental means. This trend does not hold for all other diameters as well, where, however, e.g. for 2.5 mm nails, only 75 experimental values were available. The difference between the simulated values considering a constant scatter of ±3 of the studentised residuals (**Simulation A**) and those considering two groups of residuals (**Simulation B**) is as expected. For 2.8 mm nails, the smaller scatter of small diameter nails leads to a smaller bandwidth of simulated values with higher 1st percentile and 5th percentile and a lower 99th percentile. 6 mm nails instead are already rather well represented when considering a constant scatter of ± 3 of the studentised residuals (which is no surprise, seeing that the scatter of ± 3 occurs exactly for large diameter nails).

Generally, it can be stated that the simulated databases seem to fit rather well to experimental data; in particular, if minimum and maximum simulated values are neglected. This statement does not hold when looking at the maximum experimental value for 6 mm nails (36500 Nmm), which is higher than any simulated maximum value. No mechanical or any other explanation is available as to why the experimental yield moments for 6 mm nails shown in Figure 4-18 on the right are not normally distributed, which in turn leads to this "wrong" prediction of maximum values.

To conclude, simulated databases are suitable for further analyses, which require larger datasets. Experimental gaps can be filled. However, further analyses should always not only consider simulated minimum and maximum values, but consider also simulated 1th and 99th percentiles as they may better represent experimental extreme values.

Table 4-7 Simulated and experimental yield moments in Nmm for nails with a diameter of 2.8 mm.

	Simulation A Constant variance of residuals	Simulation B Non-constant variance of residuals (d < 3.9 mm)	Simulation C Including scatter of tensile strength	Experimental values (without hdg)
n	1,000,000	1,000,000	28076	325
mean	3054	3088	3090	3294
min	1232	1906	1106	2400
max	5076	4400	5103	4400
1st percentile	2124	2485	1862	2480
5th percentile	2399	2663	2226	2598
99th percentile	3980	3688	4305	4330

Table 4-8 Simulated and experimental yield moments in Nmm for nails with a diameter of 6 mm.

	Simulation A Constant variance of residuals	Simulation B Non-constant variance of residuals ($d \geq 3.9$ mm)	Simulation C Including scatter of tensile strength	Experimental values (without hdg)
n	1,000,000	1,000,000	3047	265
mean	25181	25084	24130	26160
min	23360	22838	12338	20050
max	27204	27577	35357	36500
1st percentile	24252	23939	16145	20190
5th percentile	24527	24277	18475	22400
99th percentile	26108	26225	32336	36100

Determination of characteristic values

The discussion up to now was based on simulated values of the yield moment that were calculated using either mean or simulated tensile strength values. Furthermore, comparisons with observed experimental 5th percentiles were made. However, proper statistical determination of the 5th percentiles in accordance with EN 14358 (2016) was not considered, that take reduced sample sizes (k_s-factor) and theoretical distributions into account (usually, the lognormal distribution). The purpose of this section is to investigate if the derivation of characteristic yield moments using simulated data is valid. One possibility to calculate characteristic values is to apply Eq. (4-5) using characteristic tensile strength values and nominal diameters:

$$M_{y,Rk} = 0.1772 \cdot f_{t,k} \cdot d^{3.0191} \quad \left(\approx 1.06 \cdot \frac{1}{6} \cdot f_{t,k} \cdot d^3 \right) \tag{4-8}$$

where

$M_{y,Rk}$ Characteristic yield moment in Nmm

$f_{t,k}$ Characteristic tensile strength of nail in MPa

d Nominal diameter in mm

The expression in parenthesis in Eq. (4-8) shows the similarity to the basic theoretical equation to calculate the full plastic bending capacity of a circular cross-section, see Eq. (3-2). Therefore, to be consistent with the proposal to calculate the yield moment of staples, the suitability of Eq. (3-4), using a "characteristic yield strength $f_{y,k}$", is investigated additionally.

Table 4-9 and Table 4-10 give 5th percentiles for the yield moment of 2.8 and 6 mm nails:

- No. 1: Observed experimental 5th percentile, Figure 4-18
- No. 2: 5th percentiles of test series calculated in accordance with EN 14358
- No. 3: 5th percentile of **Simulation A** (constant variance, Table 4-7 and Table 4-8)
- No. 4: 5th percentile of **Simulation C** (including scatter of f_t, Table 4-7 and Table 4-8)
- No. 5: 5th percentile acc. to Eq. (4-8), using simulated $f_{t,k}$ = 629 MPa and 466 MPa given in Table 4-6
- No. 6: 5th percentile acc. to Eq.(3-4), using simulated $f_{t,k} = f_{y,k}$ = 629 MPa and 466 MPa given in Table 4-6
- No. 7: 5th percentiles acc. to Eq. (4-8), using $f_{t,k}$-values of test series calculated in accordance with EN 14358 (Figure 4-4 on the right)
- No. 8: 5th percentiles acc. to Eq. (3-4), using $f_{t,k} = f_{y,k}$ -values of test series calculated in accordance with EN 14358 (Figure 4-4 on the right)

Table 4-9 Different 5th percentiles of the yield moment of nails with a diameter of 2.8 mm, in Nmm.

No.	Source of 5th percentile	n	5th percentile	
1	Observed experimental value	325	2598	100%
2	Test-based values acc. to EN 14358	41	2227 to 4099	86 to 158%
3	Simulated value (constant variance)	1000000	2399	92%
4	Simulated value (including scatter of tensile strength)	28076	2226	86%
5	Value acc. to Eq. (4-8), using simulated $f_{t,k}$ = 629 MPa	1	2495	96%
6	Value acc. to Eq. (3-4), using simulated $f_{t,k}$ = 629 MPa	1	2301	89%
7	Values acc.to Eq. (4-8), using $f_{t,k}$ acc. to EN 14358	10	2442 to 3360	94 to 129%
8	Values acc. to Eq. (3-4), using $f_{t,k}$ acc. to EN 14358	10	2253 to 3099	87 to 119%

Table 4-10 Different 5th percentiles of the yield moment of nails with a diameter of 6 mm, in Nmm.

No.	Source of 5th percentile	n	5th percentile	
1	Observed experimental value	265	22400	100%
2	Test-based values acc. to EN 14358	28	18512 to 33620	83 to 150%
3	Simulated value (constant variance)	1000000	24527	109%
4	Simulated value (including scatter of tensile strength)	3047	18475	82%
5	Value acc. to Eq. (4-8), using simulated $f_{t,k}$ = 466 MPa	1	18453	82%
6	Value acc. to Eq. (3-4), using simulated $f_{t,k}$ = 466 MPa	1	16772	75%
7	Values acc.to Eq. (4-8), using $f_{t,k}$ acc. to EN 14358	11	16927 to 33426	76 to 149%
8	Values acc. to Eq. (3-4), using $f_{t,k}$ acc. to EN 14358	11	15385 to 30381	69 to 136%

The following observations can be made:

- Concerning direct evaluation of $M_{y,k}$, few individual test results were available (n = 325 and 265, No. 1) and few test series to calculate 5th percentiles in accordance with EN 14358 (n = 41 and 28, No. 2).
- For 2.8 mm nails, 5th percentiles of the tensile strength calculated in accordance with EN 14358 (needed for Nos. 7 and 8) result to 616-847 MPa, and these are based on 10 test series only. For 6 mm nails, this does not look better, and the characteristic values of $f_{t,k}$ = 427-844 MPa are based on 11 test series only. The simulated characteristic values used in Nos. 5 and 6 result to 629 MPa for 2.8 mm nails and 466 MPa for 6 mm nails, and these correspond to the respective lowest value calculated in accordance with EN 14358. This leads to the conclusion that even simulations based on few test data can deliver valid results.

- If firstly looking at experimentally based characteristic values, the large scatter gets evident. Characteristic yield moments determined in accordance with EN 14358 (No. 2), hence those yield moments which are used in design equations nowadays, range between 2442 and 3360 Nm for 2.8 mm nails and even 18512 and 33620 Nm for 6 mm nails. The minimum value of this range is lower than the observed 5th percentiles if all nails of one diameter are considered as a whole (No. 1); i.e. 2598 Nm (2.8 mm nails) and 22400 Nm (6 mm nails). General equations based on larger datasets, hence, may not capture individual low values of individuals series within that dataset.
- It could now be postulated that No. 4 delivers the most realistic characteristic yield moment, because the simulation considers both the regression residuals and the scatter of tensile strength, and hence leads to better expected values for the whole nail population. This value fits quite well to the lowest characteristic value determined in accordance with EN 14358 (No. 2). This underlines (again) the suitability of using simulated values to enlarge experimental datasets.
- Considering Eqs. (3-4) and (4-8), i.e. the mechanics-based equation to calculate the yield moment or the regression result, obviously Eq. (3-4) leads to lower yield moments for all diameters. The differences between both equations are 8 percentage points on average (Nos. 5 to 8).

No clear conclusions can be drawn from this. Hence, results for 4 mm nails are additionally given in Table 4-11, as the experimental database was largest for these nails, with 630 individual results contained in 72 test series. All in all, Eq. (3-4) is an appropriate choice, even if punishing for larger diameter nails. Eq. (3-4) allows for yield moments to be calculated based on nominal diameter and characteristic tensile strength. Figure 4-20 underlines this recommendation. On the left, characteristic yield moments are shown where the trend of Eq. (3-4) to deliver lower 5th percentiles for larger diameters, the blue line, can be seen. It gets evident that the scatter of the characteristic yield moments per diameter cannot be gathered. On the right instead, ratios are given where the black circles represent the ratio of **experimentally** determined characteristic values calculated in accordance with EN 14358 (No. 2) divided by the **simulated** characteristic values determined in accordance with Eq. (4-8) (No. 5). 13.1% of the ratios are below 1.0 (39 of 297 values) when Eq. (4-8) is used. If these ratios are calculated using Eq. (3-4) (No. 6) instead of Eq. (4-8), the red plusses, this percentage drops to 5.7% (17 of 297 values are below 1.0).

Table 4-11 Different 5th percentiles of the yield moment of nails with a diameter of 4 mm, in Nmm.

No.	Source of 5th percentile	n	5th percentile	
1	Observed experimental value	630	6270	100%
2	Test-based values acc. to EN 14358	72	5129 to 14544*	82 to 232%*
3	Simulated value (constant variance)	1000000	7539	120%
4	Simulated value (including scatter of tensile strength)	35129	6525	104%
5	Value acc. to Eq. (4-8), using simulated $f_{t,k}$ = 569 MPa	1	6622	106%
6	Value acc. to Eq. (3-4), using simulated $f_{t,k}$ = 569 MPa	1	6066	97%
7	Values acc.to Eq. (4-8), using $f_{t,k}$ acc. to EN 14358+	32	5433 to 9334	87 to 149%
8	Values acc. to Eq. (3-4), using $f_{t,k}$ acc. to EN 14358+	32	4976 to 8550	79 to 136%

* High value of 14544 Nm is from a test series with hardened nails; next largest $M_{y,k}$ is 9875 Nm.

\+ $f_{t,k}$ is ranging between 467 and 802 MPa.

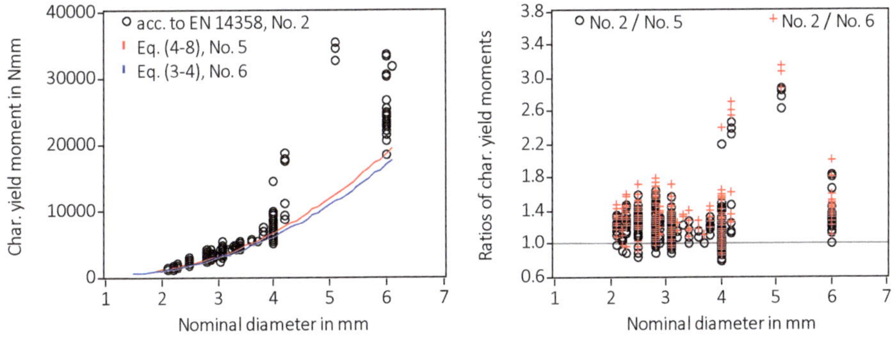

Figure 4-20 Left: Characteristic yield moments versus nominal diameters. Right: Ratios of yield moments versus nominal diameters. The given Nos. in the legend refer to the cases listed in Table 4-9 to Table 4-11.

Summary

Section 4.4.3, together with section 4.4.2, could show that:

- Data simulation techniques implemented in standard statistics software can be used to generate empirically represented data in a swift and easy-to-use way.
- Experimental datasets must be available that can provide necessary input parameters to simulate data (i.e. mean and standard deviation in case of normally distributed data), but these datasets can be small. In this case, standard deviations must be checked and, if required, higher values must be chosen.
- Databases containing simulated data can close experimental gaps and can be used for further analyses.

4.4.4 Additional test series

Additional test series in accordance with Figure 4-21 were carried out to give a closer look into the uncertainties linked to the use of different nails to determine steel properties. In series 1, "standard" certification tests were carried out using different nails to determine the yield moment M_y and the tensile capacity F_t, where the nails however were from the same batch. In series 2 and 3, both M_y and F_t were evaluated using the same nails by dividing one single nail into two parts. This procedure limited the choice of available nails, because the original nails must be long enough to provide two parts with sufficient length to be tested. Therefore, six different ringed shank nails from three producers with a length of 100 mm were chosen. The nominal diameters were 4 mm and 6 mm.

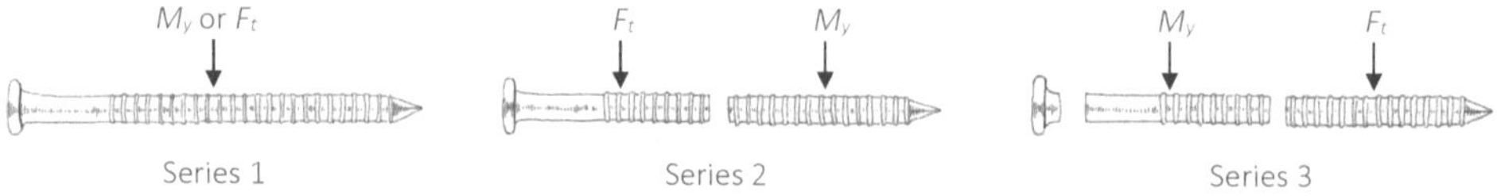

Figure 4-21 Setup of additional test series. (Drawings: Vedovelli)

Table 4-12 Results of additional test series, mean values for M_y in Nm and F_t in kN, COV in %.

	Series 1			Series 2			Series 3		
	N	$M_{y,mean}$ (COV)	$F_{t,mean}$ (COV)	N	$M_{y,mean}$ (COV)	$F_{t,mean}$ (COV)	N	$M_{y,mean}$ (COV)	$F_{t,mean}$ (COV)
d = 4 mm, all nails galvanised carbon steel									
Producer A	20	7.47 (2.0)	8.43 (2.0)	22	6.91 (1.7)	8.33 (2.2)	22	6.75 (2.7)	8.43 (2.4)
Producer B	20	7.96 (1.3)	9.30 (1.6)	22	7.98 (2.1)	9.28 (1.4)	22	8.04 (1.8)	9.10 (2.0)
Producer C	20	7.58 (1.7)	8.66 (1.5)	22	7.51 (1.6)	8.61 (1.7)	22	7.41 (2.3)	8.67 (1.4)
d = 6 mm, producer A stainless steel, producers B and C galvanised carbon steel									
Producer A	20	31.01 (1.2)	22.45 (1.2)	22	32.02 (1.7)	22.48 (1.1)	22	30.91 (1.9)	22.86 (1.2)
Producer B	20	21.16 (1.8)	15.66 (1.5)	22	20.78 (1.7)	16.10 (1.4)	22	21.93 (1.7)	15.76 (1.5)
Producer C	20	19.88 (6.9)	14.97 (2.5)	22	19.92 (5.0)	15.27 (4.5)	22	19.48 (5.7)	15.20 (5.0)

Table 4-12 gives further specifications per series and the test results in terms of mean values and coefficients of variation (COV). If looking at the steel properties, all three series delivered similar values per nail type, where the 6 mm stainless steel (1.4401) nail of producer A showed considerably higher properties than the other 6 mm nails. In series 2 and 3, the 6 mm stainless steel nails failed in the smooth shank during the tensile respectively the bending tests[67]. This was an exception to all other test results, where the failure location was within the ringed shank part. The COV is below 3% with the exception of the 6 mm nail of producer C.

For further analysis, M_y and F_t must be transferred into strength values in accordance with Eqs. (3-1) and (4-3). The measured nominal diameter d_{meas} is used in both Eqs., in analogy to section 4.4.1. Such a choice can also be motivated by looking at Figure 4-22, where the two observed failure modes of tensile tests are shown. Not all nails failed within the inner diameter d_i as can be seen on the left, but also failure surfaces ranging over inner and outer diameters were observed. Contrarily to section 4.4.1, individual test results can be compared in series 2 and 3. M_y and F_t were tested on the same nail and hence, each single yield moment can be assigned to a single tensile capacity. In series 1, the mean values must be considered for comparison of M_y and F_t; i.e. per producer and diameter, one mean value is available. Figure 4-23 on the left shows the test results in terms of tensile strength f_t and yield moment stress σ_{My} grouped by producer and diameter. Apart from the 6 mm stainless steel nails, the 6 mm nails have lower strength than the 4 mm nails, confirming the findings of Figure 4-2 on the right. Moreover, scatter within the groups is observable, including the largest scatter in group "Producer C, 6 mm", see also Table 4-12. The 6 mm stainless steel nails of producer A diverge most from the bisect line.

Giving a closer look to the scatter within one group, differences per series can be observed in Figure 4-23 on the right, where only the results for producer B are shown. The results for the 6 mm nails showed lower tensile than yield strength in series 3, whereas series 2 had values lying approximately on the bisect line. No differences in failure modes were observed between the two series, the failures always occurred in the non-smooth part of the nails. A similar clustering, albeit less clear, could also be observed for the 6 mm nails. The nails of producer A showed also a distinctive clustering between series 2 and 3, whereas the nails of Producer C did not show any differences within the series. This hints at differences along the nail shank due to production.

Figure 4-22 Tensile tests: Photos of failure modes of 6 mm carbon steel nails.

[67] Ringed shank nails made of stainless steel generally fail in the smooth shank, contrarily to ringed shank nails made of carbon steel, which fail in the non-smooth part.

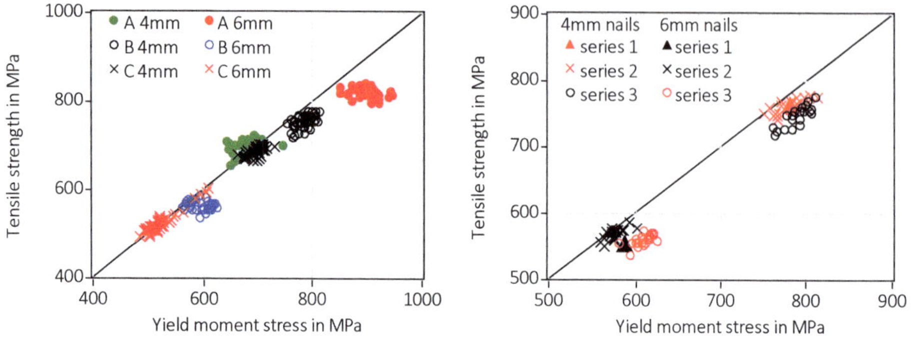

Figure 4-23 Left: Tensile strength f_t, calculated from tensile capacity F_t, versus yield moment stress σ_{My}, calculated from yield moment M_y. Results are grouped by producer and diameter. Right: Results for Producer B, differentiated by diameter and test series.

The main motivation for the additional tests was to investigate the quality of a regression in accordance with Eq. (4-4), where here, individual test values determined on the same fastener can be used. Consequently, a nonlinear regression was carried out, using all values from series 2 and 3, in total 264 test values with an individual and direct relationship between M_y and F_t. The residual analysis of the first regression resulted in 15 outliers with studentised residuals larger than $|3|$. All 15 outliers were from series 2 of the 6 mm nails of producer A (the stainless steel nails failing in the smooth shank during tensile tests), which were subsequently excluded. The regression result using the remaining 249 test values is given in Eq. (4-9) ($R^2 = 0.993$). Considering all series, i.e. including series 1 in terms of mean values, or performing a regression analysis on only mean values per series, delivered similar results with differences in the fourth decimal place. Eq. (4-9) is similar to Eq. (3-2) for the full plastic bending capacity of a circular section and the previous regression result given in Eq. (4-5).

$$M_y = 0.1562 \cdot f_t \cdot d_{meas}^{3.0624} + e \qquad (4\text{-}9)$$

where

M_y	Expected value of yield moment in Nmm
f_t	Tensile strength of nail in MPa, independent variable
d_{meas}	Mean measured nominal diameter in mm, independent variable
e	Residuals

The additional tests lead to two main conclusions. First, Eq. (4-5) is valid and the non-existing direct relationship between dependent and independent regression variables can be neglected. In fact, regressions choosing different variables do lead to results that differ by only ±10%, see also the regression based on characteristic values in Sandhaas and Görlacher (2018). Such differences vanish in the scatter of data. Second, even properties of nails from (presumably) the same batch scatter (6 mm nails of producer C) and observed COV > 5% are hence realistic. In the determination of a globally valid design equation to calculate the yield moment of nails, this real scatter, often larger than the scatter within single series, could be considered by complementing test results with simulated values.

4.5 Withdrawal

4.5.1 General

The next property to be examined is the withdrawal capacity. All withdrawal tests in the database were carried out at an angle of 90° between nail axis and grain direction. During the tests, the withdrawal capacity F_w is measured, which must be normalised in order to obtain values that are independent of the diameter and the penetration length. For dowel-type fasteners, the withdrawal parameter f_w is calculated from the withdrawal capacity F_w in accordance with EN 1382 (2016):

$$f_w = \frac{F_w}{d \cdot L_{ef}} \tag{4-10}$$

where

f_w	Withdrawal parameter in MPa
F_w	Withdrawal capacity in N
d	Nominal diameter in mm
L_{ef}	Effective penetration length in mm. L_{ef} is defined as the length of the non-smooth (profiled) part in the pointside member. That means that the tip length of the nails is subtracted from the full penetration length L_{tot}.

All tests with profiled nails were carried out with penetration lengths not larger than the profiled shank part. The timber was never predrilled. Mainly softwood was used for testing, where half of the nails per series were inserted tangentially and the other half radially to the annual rings. The insertion direction, however, was not recorded. Few test results are available for laminated veneer lumber (LVL), but these were excluded in the following analyses unless otherwise stated[68]. A histogram of the softwood densities is given in Figure 4-24 on the left, where the influence of different standards valid at the time of testing can be seen. The considerable amount of timber with densities around 350 kg/m³ for instance is due to the 1989 version of ISO 8970 (1989), where the sampling rules required that the densities of the timber used for testing should lie around a characteristic value of 350 kg/m³. The testing procedure, however, did not change. Furthermore, the coefficient of variation (COV)

[68] Concerning LVL, 40 tests with softwood LVL with ρ_{mean} = 602 kg/m³ (COV = 0.02) were carried out, and 20 tests with beech LVL with ρ_{mean} = 771 kg/m³ (COV = 0.07). The LVL was not predrilled.

of the density of most test series was rather low. This can be seen in Figure 4-24 on the right, where the observed coefficients of variation of the density that vary mainly between 0.03 and 0.06. Only mean densities were recorded for 34 test series leading to a COV of zero.

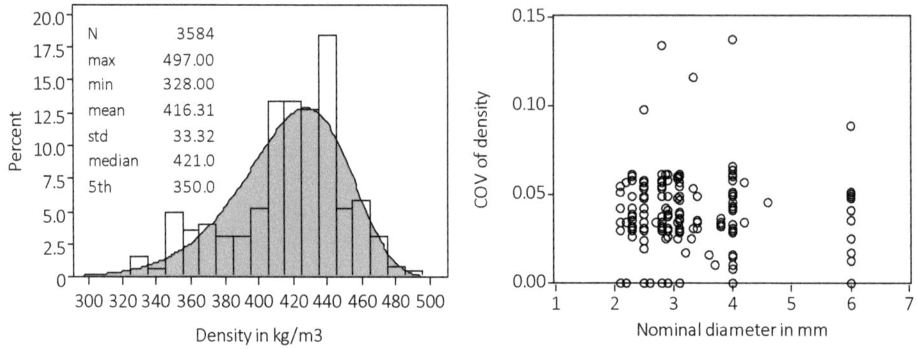

Figure 4-24 Left: Histogram with fitted Weibull distribution of densities of timber used for withdrawal tests. Right: Observed coefficient of variation of densities, 181 test series.

In current design rules, the withdrawal parameter f_w of smooth shank nails is a function of the density, see Eq. (4-2). Considering the values in the database, a new regression can be carried out with the withdrawal capacity (or thereof derived values such as f_w) as dependent variable, and the following independent variables: Measured, nominal, inner and outer diameters, penetration length with and without tip length and the timber density. Similar to the previous analyses concerning the yield moment, the measured and not the nominal diameter could be used, e.g. to calculate f_w. Figure 4-25 on the left shows the differences between the withdrawal parameters calculated in accordance with Eq. (4-10), once with the nominal and once with the measured diameter, where a maximum difference of 8% is observed. However, seeing the maximum difference of 8% and the usually large scatter of test results, the first investigation step is to identify those variables that determine the withdrawal capacity of a nail. A further reason to give a closer look to the database is Eq. (4-10), which is only a vehicle to normalise withdrawal capacities and can be considered being a model itself. The denominator is not calculating the surface area (but the projected area) and an even stress distribution along the nail shank is assumed. The full penetration length varied between $7.3 \cdot d$ and $20 \cdot d$ for nails with a diameter of up to 4 mm and between $8 \cdot d$ and $13.3 \cdot d$ for 6 mm nails. It can be assumed that the stress distribution along the shank will be more uneven for higher penetration lengths, and that in such cases Eq. (4-10) is not fully correct. This simplification also motivates the choice to use the projected area $d \cdot L_{ef}$ and not the surface area $\pi \cdot d \cdot L_{ef}$, which is mechanically more correct. As the difference in f_w-values would be a constant, namely π, the only important rule is to be consistent[69].

[69] See also more in-depth discussion in section 5.5.2.

As a first step, the correlation between the withdrawal parameter f_w (in accordance with Eq. (4-10), using the nominal diameter) and the density ρ is analysed resulting in a Pearson's correlation coefficient of R = 0.32. In Figure 4-25 on the right, the withdrawal parameter is shown in dependence of the density, and the shown trendline shows the correlation between f_w and ρ. The differences between the individual nails are obviously much higher than the differences caused by the density, as the scatter of the results per density is very high. These differences between the individual nails may be caused by different production qualities (new or worn tools) and from wood characteristics other than the density. In Figure 4-26, 2416[70] individual withdrawal parameters are given per test series versus increasing nominal diameters. The scatter of the densities used within the individual series was similar for small and large diameter nails as can be seen in Figure 4-24 on the right. The upper envelope line illustrates the lower scatter for 4 to 6 mm nails compared to nails with smaller diameters, which can reach higher individual withdrawal parameters. The lower values for f_w remain instead similar for all diameters indicating a negative correlation between f_w and d (indeed, R = -0.19). Different hypotheses may explain this dependence of f_w on d.

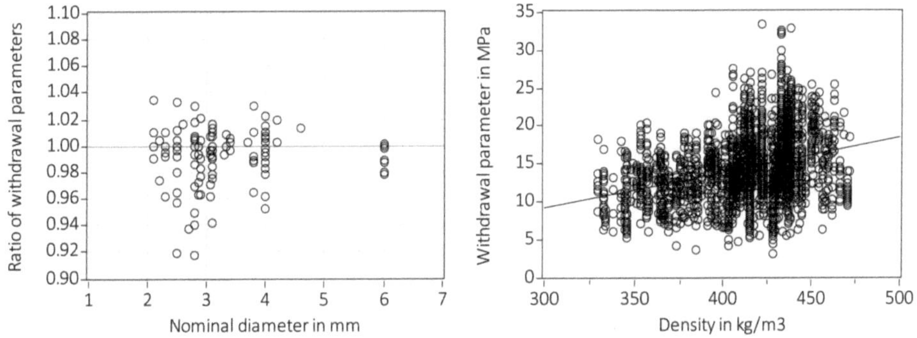

Figure 4-25 Left: Ratio of withdrawal parameters: Parameter calculated with measured diameter divided by parameter calculated with nominal diameter, 2416 values. Right: Withdrawal parameter f_w versus density ρ, trendline line is shown, 2416 values.

[70] The smaller number compared to Table 4-1 is due to the fact that only test results were included where the tip length was measured. Furthermore, test results using beech and softwood LVL were excluded.

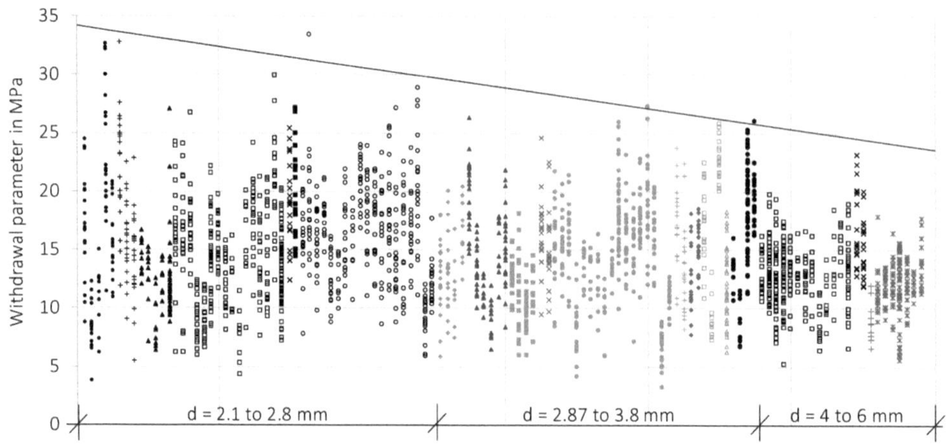

Figure 4-26 2316 withdrawal parameters per test series versus increasing nominal diameters. Each vertical line comprises the individual test results per series, the different symbols identify the different nominal diameters. An upper envelope line is shown.

One reason for the larger scatter observed for small diameter nails could be that these nails are inserted in either early- or latewood. This would lead to high individual withdrawal capacities if the nail is inserted and withdrawn from latewood. For large diameter nails instead, the evaluated withdrawal capacity is a smeared value as these nails are inserted in both early- and latewood. This hypothesis highlights a dependence of f_w on wood characteristics that cannot be fully explained by the (global) density. Another hypothesis concerns production issues. In Figure 4-1 on the left, a ringed shank nail is shown, where different sharpness of the rings can be seen. The rings closer to the nail head have less pronounced rings than the rings closer to the tip. During testing, it has been observed that the sharper the rings, which can be felt when passing the nails through the fingers, the higher is the measured withdrawal capacity. The quality of the rings will scatter above all between nails and not only within one nail. Sharp rings will be produced when new tools are used and less sharp rings when the tools are worn. This, in turn, leads to different f_w-values in dependence of the age of the tools used to produce the non-smooth shank nails. This quality issue of non-smooth shank rings was already part of a detailed study by White and Gales (1990), who quantified the variation of thread qualities and their influence on withdrawal resistance. Another possible influencing factor on the withdrawal parameter could be the ring height, i.e. the relation between the inner (d_i) and the outer (d_o) diameter, see Figure 4-1 on the right. This assumption is however not supported by the correlation coefficient (R = 0.0) between d_i/d_o and f_w. The scatter of test results is indeed very high as can be seen in Figure 4-27 on the right, which shows the standard deviations of the 119 test series (assuming a lognormal distribution). Not a single standard deviation is below 0.07 with 0.36 being the highest value.

It can be concluded that a regression analysis will not lead to meaningful equations to calculate withdrawal parameters from e.g. the density, as other factors have a strong influence on experimentally determined withdrawal capacities. At least at the current state of production technology, it will be difficult to guarantee uniform sharpness of rings and spirals of non-smooth shank nails. Additionally, differences in withdrawal capacities due to small or large nail diameters will remain.

4.5.2 Influence of tip length

Up to now, the withdrawal parameter f_w was calculated according to Eq. (4-10), where, per definition, the tip length of the nails was subtracted to define an effective penetration length. This is the scientifically sound approach, as the tip length does not contribute to the withdrawal capacity. For joint design purposes however, the calculation of the withdrawal capacity F_w with the effective penetration length L_{ef} is rather cumbersome for practitioners, as they have to deal with different lengths for lateral and axial loads. Furthermore, the tip length is usually unknown. Additionally, considering the high scatter shown in Figure 4-26 and Figure 4-27 on the right, the necessity of using L_{ef} instead of the full penetration length L_{tot} remains worth discussing; i.e. subtracting the tip length or not may not influence withdrawal parameters significantly. This is more thoroughly investigated in the following. The tip length L_p itself is a function of the nominal diameter. This is shown in Figure 4-27 on the left, where the length of the tip varies between d and $2 \cdot d$.

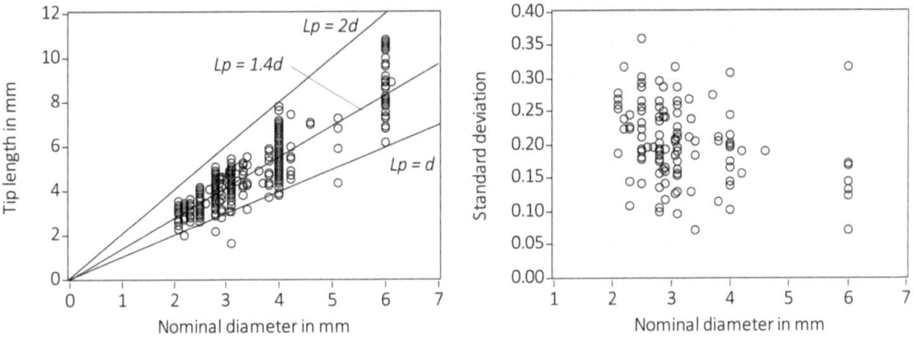

Figure 4-27 Left: Tip length L_p versus nominal diameter d, 5613 values. Right: Standard deviation of withdrawal parameters (lognormal distribution), 119 series.

In principle, F_w can be calculated using the effective penetration length L_{ef}, **Method *A***, or the full penetration length L_{tot}, **Method *B***. Considering now that $L_p = 1.4 \cdot d$ (Figure 4-27 on the left), the withdrawal parameters can be derived from the test results as follows, where the superscripts indicate whether **Method *A*** or ***B*** is used.

$$f_w^A = \frac{F_w^{test}}{d \cdot L_{ef}} = \frac{F_w^{test}}{d \cdot (L_{tot} - 1.4 \cdot d)} \quad \text{(a)}$$

$$f_w^B = \frac{F_w^{test}}{d \cdot L_{tot}} = \quad \text{(b)} \qquad \text{(4-11)}$$

For the test series included in the database, full penetration lengths L_{tot} between $7.3 \cdot d$ and $20 \cdot d$ were used with a mean value of $11.3 \cdot d$ (COV = 23%). Assuming now that the withdrawal tests were performed with a full penetration length of $12 \cdot d$, the ratio of both withdrawal parameters given in Eq. (4-11) can be calculated:

$$\frac{f_w^B}{f_w^A} = \frac{d \cdot (L_{tot} - 1.4 \cdot d)}{d \cdot L_{tot}} = \frac{12 \cdot d - 1.4 \cdot d}{12 \cdot d} = 0.883 \qquad \text{(4-12)}$$

The design equations to calculate the withdrawal capacities are:

$$F_w^A = f_w^A \cdot d \cdot L_{ef} = f_w^A \cdot d \cdot (L_{tot} - 1.4 \cdot d) \quad \text{(a)}$$

$$F_w^B = f_w^B \cdot d \cdot L_{tot} \quad \text{(b)} \qquad \text{(4-13)}$$

Consequently, the difference between **Methods** *A* and *B* can be expressed by:

$$\frac{F_w^B}{F_w^A} = \frac{f_w^B}{f_w^A} \cdot \frac{d \cdot L_{tot}}{d \cdot (L_{tot} - 1.4 \cdot d)} = 0.883 \cdot \frac{1}{1 - 1.4 \cdot \frac{d}{L_{tot}}} \qquad \text{(4-14)}$$

This difference between both methods in dependence on the slenderness L_{tot}/d is shown in Figure 4-28 on the left, where practical lower and upper limits of the penetration lengths are chosen. Penetration lengths above $20 \cdot d$ are not realistic, as nail tensile failure may occur. The maximum overestimation of F_w when including the tip length is 7% at a full penetration length of $8 \cdot d$ and the respective underestimation at $20 \cdot d$ is 5%. This is a negligibly small difference considering the scatter in the test results. Therefore, the subtraction of the tip length is not necessary, and F_w can be calculated considering the full penetration length L_{tot}, **Method** *B*.

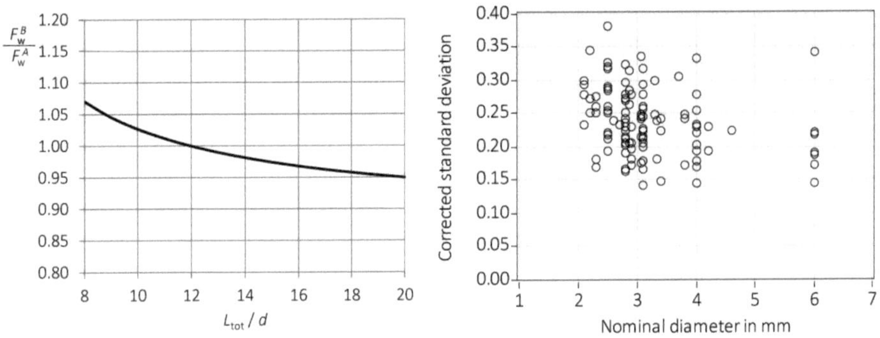

Figure 4-28 Left: Difference between withdrawal capacities calculated with **Method *A*** and ***B***, Eq. (4-14), in dependence of the slenderness L_{tot}/d. Right: Corrected standard deviations of (logarithmic) withdrawal parameter in accordance with Eq. (4-17), 119 test series.

To further look into this issue, a small test series was carried out, where 4 mm ringed shank nails were withdrawn from five pieces of timber, inserting eleven nails per timber piece with eleven different penetration lengths along the fibre direction[71]. The densities of the timber pieces were 375, 388, 437, 452 and 456 kg/m^3 at moisture contents between 12.5% and 13.4%. The chosen penetration lengths were {2, 3, 4, 6, 8, 10, 12, 14, 16, 18, 20} · nominal diameter *d*. This means that per penetration length, five test results are available. Currently, the minimum penetration length in accordance with Eurocode 5 is 6 · *d* for laterally and axially loaded non-smooth shank nails, where however the withdrawal capacity must be reduced for penetration lengths smaller than 8 · *d* for axially loaded nails. No maximum value is given. The geometric data of the nails were measured, and the measured tip length of the nail was ca. 4.3 mm, i.e. $L_p \approx d$, see also Figure 4-29 on the left. The test results in kN are given in Table 4-13. As the nails used for the additional tests in section 4.4.4 were used also here, their tensile capacity is known and its mean value was $F_{t,mean}$ = 8.64 kN. This is twice the maximum withdrawal capacity $F_{w,max}$ = 4.35 kN observed for the maximum penetration length of 20 · *d* in the timber piece with the highest density of 456 kg/m^3.

[71] Spruce solid timber was used and all nails were inserted in the radial direction. The timber was not predrilled. All nails were made of carbon steel and galvanised and were from the same batch, i.e. the quality of their rings should be comparable. Per timber piece, the eleven nails were perfectly aligned along the grain in order to reduce the influence of the timber on the test results as much as possible; i.e. so that per timber piece, the penetration length was the "only" varied parameter.

Table 4-13 Results of additional withdrawal tests on ringed shank nails, F_w in kN.

ρ in kg/m³	m.c. in %	Penetration length, nominal diameter d = 4 mm $2 \cdot d$	$3 \cdot d$	$4 \cdot d$	$6 \cdot d$	$8 \cdot d$	$10 \cdot d$	$12 \cdot d$	$14 \cdot d$	$16 \cdot d$	$18 \cdot d$	$20 \cdot d$
375	13.4	0.16	0.31	0.42	0.89	1.32	1.69	2.06	2.29	3.16	3.68	3.88
388	12.9	0.16	0.34	0.57	0.90	1.26	1.71	2.14	2.53	2.82	3.53	3.53
437	12.8	0.14	0.21	0.35	0.77	1.30	1.58	1.75	2.08	2.65	3.08	3.36
452	12.6	0.12	0.17	0.36	0.73	1.18	1.70	1.81	2.37	2.42	2.58	3.74
456	12.5	0.18	0.38	0.58	1.21	1.53	2.41	2.67	3.37	4.18	4.09	4.35

The determined withdrawal capacities given in Table 4-13 were normalised applying Eq. (4-10), i.e. subtracting the tip length from the penetration length ($f_{w,Lef}$). The capacities were also normalised applying Eq. (4-11), i.e. using the full penetration length ($f_{w,Ltot}$). The measured nominal diameter of d_{meas} = 3.96 mm was used for this normalisation. Figure 4-29 on the right shows the test results as ratio $f_{w,Ltot}$ over $f_{w,Lef}$ versus the penetration length. Unsurprisingly, the influence of the tip is clearly present for small penetration lengths, and this trend is due to the normalisation method[72]. Nevertheless, what can be said is that the difference in withdrawal parameters is 8% between a penetration length of $8 \cdot d$ and that of $20 \cdot d$. This clearly motivates the choices concerning minimum penetration lengths in the current Eurocode 5 – if the tips of all nails have always more or less the same length in relation to the nail diameter.

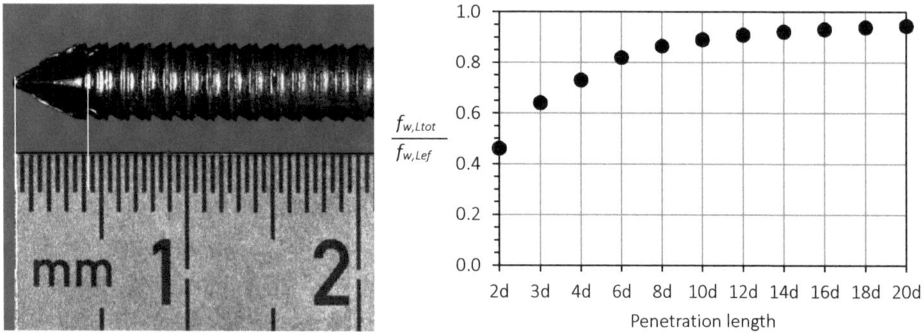

Figure 4-29 Left: Detail of tip area of used nails with measured nominal diameter d_{meas} = 3.96 mm. Right: Ratio of $f_{w,Ltot}$ (including tip) over $f_{w,Lef}$ (excluding tip) versus full penetration length.

[72] This is trivial mathematically; the tip length is constant and hence, the longer the penetration length the smaller the percentage of the tip on the inserted length. In other words, the ratio of tip length over penetration length is the same as the ratio shown in Figure 4-29. This difference will always be present; and values $f_{w,Lef}$ will always differ from $f_{w,Ltot}$ by this percentage.

The main question is, however, if the tip has an influence on the withdrawal parameter, i.e. not only purely due to the mathematical definitions. If now representing the test results in terms of withdrawal parameter versus the penetration length as done in Figure 4-30, more conclusions can be drawn. If looking at Figure 4-30 on the left, the timber pieces with high densities seem to oscillate more and lower densities seem to provide more uniform results. This may again be due to other wood characteristics than the density alone, similar to the discussion around Figure 4-26. Figure 4-30 on the right shows the same data, but as a box-plot grouping the five results per penetration length. The results show a similar asymptotic trend as Figure 4-29 on the right, but now in terms of withdrawal parameters and not as (mathematical) ratios. This asymptotic trend means that the tip length is influencing the results until up to a penetration length of approximately $8 \cdot d$, with similar mean values for all penetration lengths $> 8 \cdot d$. This is also confirmed when looking at Figure 4-30 on the bottom, as even when subtracting the tip length from the penetration length, mean withdrawal parameters are increasing at smaller penetration lengths[73]. If neglecting the results for a penetration length of $2 \cdot d$ in Figure 4-30 on the bottom right, $f_{w,Ltot}$ and $f_{w,Lef}$ show a similar trend.

[73] The test results at a penetration length of $2 \cdot d$ are different; showing a high mean value and less scatter.

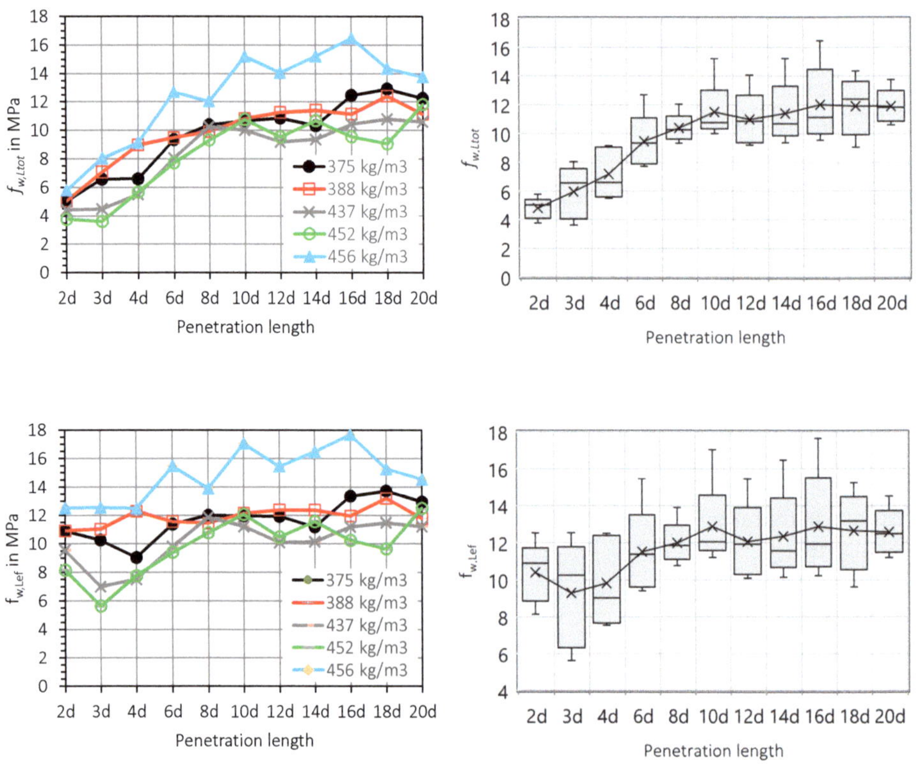

Figure 4-30 Withdrawal parameter versus the full penetration length. Left: Individual values per series. Right: As boxplot with curve through mean values. Top: Withdrawal parameter $f_{w,Ltot}$ in accordance with Eq. (4-11). Bottom: Withdrawal parameter $f_{w,Lef}$ in accordance with Eq. (4-10).

4.5.3 Characteristic values

Independently of any problems arising from the large scatter of test results, characteristic values for the withdrawal parameter are needed for current joint design. The coefficients of variation of the two main scattering properties, withdrawal parameter and density, are different, see Figure 4-24 on the right and Figure 4-27 on the right. These differences in scatter must be considered when calculating 5th percentiles. For softwood, coefficients of variation (COV) of the density amount to ca. 10% (Colling, 1990, p. 86), and the observed coefficients of variation, Figure 4-24 on the right, are mainly lower. Consequently, the withdrawal parameters were determined on samples that were not representative. In the following, the procedure of Annex D of prEN 14592 (2017) was applied to determine corrected coefficients of correlation and hence characteristic values of the withdrawal parameters.

As a first step, the dependency of the withdrawal parameter f_w on the density ρ must be determined. Although the scatter of results is large, a weak correlation between f_w and ρ

exists (R = 0.32), see Figure 4-25 on the right. Therefore, a nonlinear regression is carried out, using the density as the only explanatory variable. Differently to Eq. (4-2), the exponent of the density is not forced to be quadratic:

$$f_w = 4.2 \cdot 10^3 \cdot \rho^{1.35} \tag{4-15}$$

where

f_w Withdrawal parameter in MPa, calculated in accordance with Eq. (4-10)

ρ Density in kg/m^3

Eq. (4-15) is based on 2416 test results, where the tip length was measured and hence f_w could be calculated. The exponent of 1.35 corresponds well to the correction factors proposed in prEN 14592 (2017, Table D.1), where the regression is however rather weak with $R^2 = 0.10$. The exponent of 1.35 is used to correct the withdrawal parameters of each test series (i.e. to obtain a horizontal regression line in Figure 4-25 on the right), using a reference mean density of 420 kg/m^3 (= ρ_{mean} of C24, EN 338, 2016):

$$f_{w,corr} = f_w \cdot \left(\frac{420}{\rho_{mean}} \right)^{1.35} \tag{4-16}$$

where

$f_{w,corr}$ Corrected individual withdrawal parameter in MPa

f_w Individual withdrawal parameter in MPa, calculated in accordance with Eq. (4-10)

ρ_{mean} Mean density of each test series in kg/m^3

The corrected withdrawal parameters $f_{w,corr}$ are assumed to have a lognormal distribution. A normal distribution is assumed for the density. Using the approach given in Annex D of prEN 14592, the standard deviations of the corrected withdrawal parameters can now be adjusted so that they reflect the timber population, where COV = 0.10:

$$std_{fw,corr} = \sqrt{std_{fw}^2 + 1.35^2 \cdot \left(0.10^2 - COV_\rho^2\right)} \tag{4-17}$$

where

$std_{fw,corr}$ Corrected standard deviation of withdrawal parameter, per test series

std_{fw} Observed standard deviation of withdrawal parameter, per test series

1.35 Correction factor, see Eq. (4-15)

0.10 COV of density of timber population

COV_ρ Observed COV of density, per test series

Figure 4-28 on the right shows the corrected standard deviations, which are higher than the uncorrected standard deviation based on a lognormal distribution, Figure 4-27 on the right.

Finally, in accordance with EN 14358 (2016), 5th percentile values for $f_{w,corr}$ were estimated assuming a lognormal distribution and using the corrected standard deviations $std_{fw,corr}$. The limited amount of test results per test series was considered applying the k_s-factor given in EN 14358. Figure 4-31 shows the characteristic withdrawal parameter $f_{w,corr,k}$ per test series versus the nominal diameter d. The five nail types are identified. Spiral, special spiral and square nails did not reach characteristic values larger than 8 MPa. The validity of the evaluated characteristic values must be discussed however. For instance, only values for ringed shank nails are available for 2.1 mm nails, and the determined characteristic values $f_{w,corr,k}$ range from 5.0 MPa to 12.3 MPa. Considering that the general ring shape (with $d_i = 0.86 \cdot d_o$) does not change significantly between the nails (five different test series and four different producers), there is no straightforward explanation why the results differ by a factor of 2.5, except for the hypotheses formulated above. This randomness of the characteristic values or, in other words, the dependency of the withdrawal parameters on influencing factors not measured within the framework of certification tests, is further highlighted, when the horizontal lines in Figure 4-31 are considered. These lines correspond to some withdrawal classes for all fastener types in accordance with prEN 14592 (2017, Table 4). The value of 12 MPa corresponds to the highest value for self-tapping screws, and it is questionable if such high values can be reliably reached with ringed shank nails.

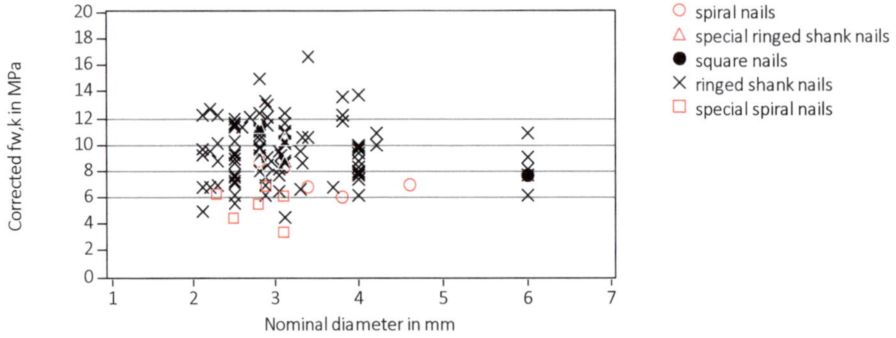

Figure 4-31 Characteristic withdrawal parameters $f_{w,corr,k}$ versus nominal diameter, 119 test series.

It can be questioned why ρ_{mean} per test series is used in Eq. (4-16) to correct individual withdrawal parameters instead of the individual densities ρ per test. If re-plotting Figure 4-31 with $f_{w,k}$-values based on $f_{w,corr}$-values corrected with individual densities, however, differences are minor (no difference on the average). Only two series, with higher coefficients of variation of the density of 10% and 12% (see Figure 4-24 on the right), showed differences of 12% resp. 14%.

Based on the actual database and the analyses shown, it can be concluded that also in future, tests have to be carried out and values for $f_{w,k}$ have to be taken from technical documents. With regard to code implementation, technical classes could be introduced so that designers do not need to consult technical documentation to get withdrawal parameters. The introduction of upper limits should be considered to address production issues (e.g., $f_{w,k}$ is derived using nails produced in new machines and subsequent production is then done with worn machines leading to less pronounced rings).

4.6 Head pull-through

A total of 68 test series are available that contain 1039 individual test results (mostly 10 or 20 tests per series). The head pull-through capacity F_{head} corresponded to the maximum force recorded before or at a machine displacement of 15 mm (see also section 5.6). The ratio of the head diameter d_h over the nominal diameter d of the analysed nails was greater than 1.8 in all cases, with a mean value of $d_h = 2.3 \cdot d$ (see Figure 4-32 on the left). The head pull-through parameter f_{head} is calculated from the head pull-through capacity F_{head} in accordance with EN 1383 (2016):

$$f_{head} = \frac{F_{head}}{d_h^2} \tag{4-18}$$

where

f_{head} Head pull-through parameter in MPa

F_{head} Head pull-through capacity in N

d_h Mean measured head diameter of each test series in mm

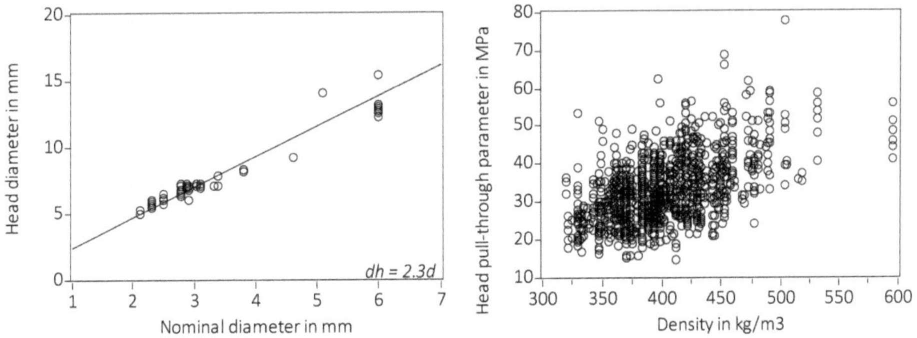

Figure 4-32 Left: Head diameter d_h versus nominal diameter d, 1039 values. Right: Head pull-through parameter f_{head} versus density ρ, 1039 values.

Similar to the withdrawal parameter, also the head pull-through parameter f_{head} is considered to be a function of the timber density ρ, see Eq. (4-2)[74]. The scatter is again high, see Figure 4-32 on the right, where however Pearson's correlation coefficient is higher with R = 0.52 (withdrawal: R = 0.32). The head shape may have an additional influence, but only round, flat heads are included in the database[75]. No head pull-through tests were carried out on nails with trumpet heads[76] as these nails are used to fasten steel plates to timber members, and no head pull-through parameters are required. Only solid wood made of softwood (usually spruce, all not predrilled) was used. The data with densities $\rho > 590$ kg/m³ was deleted for all further analyses, because the densities are improbable for spruce and exert a strong influence on regression results. The timber thickness was not given for 600 tests. The other 439 tests were carried out with thicknesses between $6 \cdot d$ and $16 \cdot d$.

In the following, analyses similar to Section 3.4 are discussed. Analogously to the withdrawal parameter, characteristic values per test series are calculated and analysed, where both the head pull-through parameter and the coefficients of variation are corrected. A nonlinear regression using 1031 individual test results resulted in ($R^2 = 0.27$):

$$f_{head} = 11.2 \cdot 10^{-3} \cdot \rho^{1.33} \tag{4-19}$$

where

f_{head} Head pull-through parameter in MPa, calculated in accordance with Eq. (4-18)

ρ Density in kg/m³

The exponent of 1.33 is used to correct the head pull-through parameters of each test series (i.e. to obtain a horizontal regression line in Figure 4-32 on the right), using a reference mean density of 420 kg/m³ (= ρ_{mean} of C24, EN 338, 2016):

$$f_{head,corr} = f_{head} \cdot \left(\frac{420}{\rho_{mean}} \right)^{1.33} \tag{4-20}$$

where

$f_{head,corr}$ Corrected individual head pull-through parameter in MPa

f_{head} Individual head pull-through parameter in MPa, in accordance with Eq. (4-18)

ρ_{mean} Mean density of each test series in kg/m³

---

74 Similar to the withdrawal parameter, the head pull-through parameter is a conventional value, because the normalisation in accordance with Eq. (4-18) does not consider the stressed area (which would be round in the case of a round head). Both values are indeed parameters and no strength values. See sections 5.5.2 and 5.6 for further discussions.

75 The exact head shape is never measured in the reports nor are always photos or shop drawings available. Therefore, mixed shapes between flat and countersunk heads may exist.

76 A ringed shank nail with a trumpet head is shown in Figure 4-1 on the left on top. Generally, these nails have nominal diameters of 4 mm and 6 mm.

The corrected head pull-through parameters $f_{head,corr}$ are assumed to have a lognormal distribution and the logarithm is taken. A normal distribution is assumed for the density. Using the approach given in Annex D of prEN 14592, the standard deviations of the corrected head pull-through parameter can now be adjusted so that they reflect the timber population. This is done in analogy to Eq. (4-17), again assuming a COV of 10% for the density of the timber population.

Figure 4-33 on the left shows the coefficients of variation of the density and the corrected standard deviations std_{corr} of the head pull-through parameter for each of the 68 test series. The variation of the density within individual test series was higher than for the withdrawal tests, where most coefficients of variation were below 0.06, see Figure 4-24 on the left. For the sake of completeness, Figure 4-33 on the right shows the histogram of densities.

Finally, in accordance with EN 14358 (2016), 5th percentile values for $f_{head,corr}$ were estimated assuming a lognormal distribution and using the corrected standard deviations std_{corr}. The limited amount of test results per test series was considered applying the k_s-factor given in EN 14358. Figure 4-34 shows the characteristic head pull-through parameter $f_{head,corr,k}$ per test series versus the nominal diameter d. The different nail types are identified where the nomination "smooth shank" identifies those non-smooth shank nails where the smooth part of the shank underneath the nail head was longer than the timber thickness through which the nail was pulled. It seems that a smooth shank underneath the head reduces the head pull-through parameter systematically. This explains the lower head pull-through values of the 6 mm nails. Referring back to the head shape (see footnote 75), the series with $f_{head,corr,k}$ = 33.5 MPa is a special nail with a flat head. The nails of the series with $f_{head,corr,k}$ = 32.4 MPa are available in two head shapes, a flat head and a head with a slight countersunk of 25°, and it is not clear which version was used during testing. No information on the head shape is given for the ring shanked nail with $f_{head,corr,k}$ = 17.9 MPa.

Based on the database, a head pull-through parameter of 15 MPa in timber with a density ρ_k = 350 kg/m^3 (observed 5th percentile of database was 338 kg/m^3, Figure 4-33 on the right) could be used without further testing, for all nails with non-smooth shanks as long as d_h/d > 1.8. This value is significantly higher than the parameter of 8.6 MPa for smooth shank nails (see Eq. (4-2) with ρ_k = 350 kg/m^3), because the head pull-through parameter of non-smooth shank nails includes the withdrawal capacity of the shank[77] (see also Eurocode 5, 2010, Eqs. 8.23 and 8.24). Depending on the shape of the shank underneath the head, higher values are possible, but these values have to be determined by testing and declared in technical documentation for the individual nails. Generally, the head pull-through parameter is determined with only one specific nail length. As this value is assigned to other nail lengths as well, it has to be ensured that the tests are performed with the most unfavourable shank under the head, which can occur (i.e. the highest percentage of smooth shank).

---

[77] This is different to F_{head}-values for screws, where only the system "screw head + smooth shank" is tested.

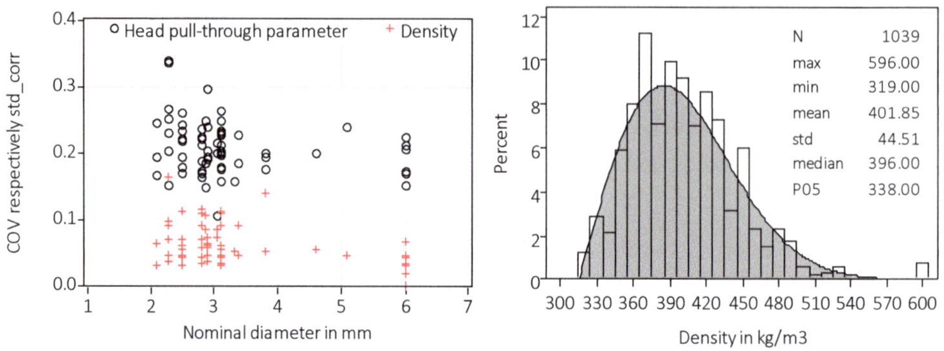

Figure 4-33 Left: Observed coefficients of variation of the density and corrected standard deviations of the (logarithmic) head pull-through parameter, 68 test series. Right: Histogram with fitted Weibull distribution of densities of timber used for head pull-through tests; high densities included.

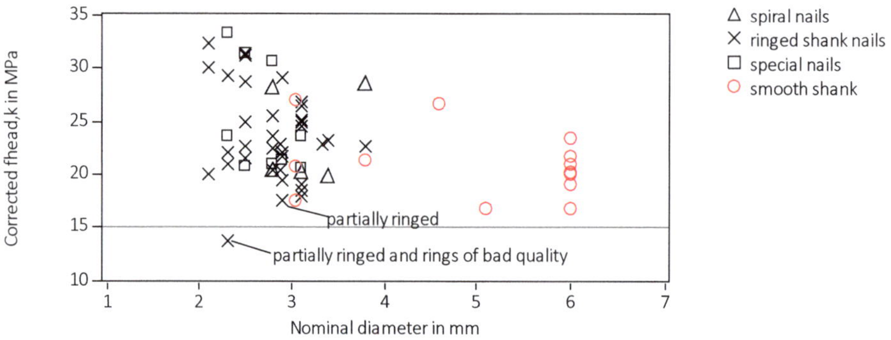

Figure 4-34 Characteristic head pull-through parameters $f_{head,corr,k}$ versus nominal diameter, 68 test series. "Smooth shank" means that the shank inserted in the timber was smooth although non-smooth shank nails were tested. "Partially ringed" means that the shank inserted in the timber was partially smooth and partially non-smooth.

With regard to code implementation, technical classes could be introduced for head pull-through parameters similar to withdrawal classes. Considering the persistent scatter of $f_{head,k}$ in Figure 4-34 for similar head shapes (round, flat shape and d_h approx. $2 \cdot d$), the random selection of the used timber seems to have a significant influence. Parameters such as annual ring widths and orientation of tangential and radial directions may impact on the experimental values.

4.7 Conclusions

A comprehensive database containing test results on mainly ringed shank nails was analysed. The following concrete consequences can be considered concerning input parameters for joint design in accordance with Eurocode 5 (DIN EN 1995 1-1, 2010):

- *Wire tensile strength*: These tests are not needed. However, producers may still require wire tests to check delivered steel grades.
- *Nail tensile capacity*: Tensile tests on finished nails need to be carried out and subsequently, a nail tensile strength f_t can be calculated using the nominal diameter. This tensile strength could be assigned to different technical classes, depending on the nail diameter and/or the steel grade resp. finishing. Furthermore, the present Eurocode 5 does not take the nail tensile capacity F_t into account. For large penetration lengths, high timber densities and high head pull-through values however, a tensile failure may be governing. Therefore, it is recommended either to include rules for verifying the tensile strength or to limit the accountable penetration length to e.g. $20 \cdot d$. It is recommended to prescribe a larger maximum standard deviation than the value of 0.05 given in EN 14358 (2016). This is due to the fact that scatter between batches is higher than within batches, and nevertheless test results for nails from one batch are extrapolated to nails from different batches.
- *Yield moment*: The equation defining the theoretical full plastic bending capacity of a round section, Eq. (3-2), can be inserted in Eurocode 5 for all nail types (with round cross-sections), using nominal diameter and characteristic nail tensile strength to calculate the characteristic yield moment. The characteristic tensile strength must be taken from technical documentation for the individual nails or technical classes can be used to define different characteristic tensile strength values. Depending on how many classes shall be defined, predicted $M_{y,k}$-values are more or less conservative. These may lead to wrong predictions of the failure modes; i.e. a failure with two plastic hinges is predicted, but the higher M_y of the nail results in rigid behaviour of the fastener which does not develop a plastic hinge. Therefore, also upper bound strength values should be defined for cases where it is essential that failure modes are correctly predicted. It is recalled that the value at a bending angle of 45° or at rupture is considered here as M_y.
- *Withdrawal parameter*: Considering the analysed database with its limitations, no design equations are possible. It could be shown that other properties than timber density have a considerable influence on the withdrawal capacity. Above all, the quality and sharpness of the rings significantly determine the withdrawal capacity. Technical classes could be introduced, imposing, however, an upper bound value to address production issues.

- ***Head pull-through parameter***: The conclusions are similar to those for the withdrawal parameter. However, similar to current regulations for screws which allow for a lower-bound value for $f_{head,k}$ (see footnote 143), a minimum head pull-through parameter of 15 MPa in timber with a characteristic density of ρ_k = 350 kg/m^3 could be used without further testing, for all nails with non-smooth shanks as long as $d_h/d > 1.8$. Of major importance for head pull-through are the characteristics of the shank part inserted in the timber through which the nail is pulled. If only smooth shank parts are inserted in the timber, then the head pull-through capacity will be smaller. This has to be taken into account when choosing the representative nails for testing.

All properties show a considerable scatter whose sources could not be completely identified and only hypotheses could be developed that could explain the variation. Data on possible influencing factors were lacking or cannot be measured, such as nail production quality or local wood properties directly around the nails. Consequently, statistical analyses suffer from incomplete data. This is a general issue of retrospectively assembled databases where datasets are unequally distributed, e.g. less results for larger diameters, or information is missing, e.g. sharpness of rings. However, the potential of using simulated data for boosting datasets could be shown, which could help to overcome some of the challenges.

5 Screws

5.1 State-of-the-art

Self-tapping timber screws are one of the most important fastener types in modern timber engineering. Due to their good performance, ease of application and versatile ranges of use, the development of modern timber screws is one of the primary factors for the advent of many modern engineered timber structures. Screws are often designed to accommodate specific purposes, such as screws optimised for specific timber products (Brandner, 2019) or screws with variable thread geometries to pre-stress timber (Steilner, 2014). Furthermore, screws are an effective means to reinforce against tensile failures perpendicular to the grain (Bejtka, 2003; Bejtka and Blaß, 2005), shear (Dietsch, 2012) or as reinforcement of beam supports (Bejtka and Blaß, 2006). Generally, screws in joints can be loaded perpendicular to their axis ("joints with laterally loaded screws") or in direction of their axis ("joints with axially loaded screws").

In general, the European Yield Model can be used to design joints with laterally loaded screws (see section 2.5). The speed of screw development is so immense that the current Eurocode 5 (EN 1995 1-1, 2010) does not include bespoke rules for self-tapping screws, with the exception of few specific rules concerning axially loaded screws. Indeed, in order to design joints with screws, most input parameters (e.g. tensile capacity, yield moment) must be taken from technical documentation of the screws. In Europe, self-tapping screws can be certified in accordance with EN 14592 (2012) or through a European Technical Assessment (ETA) based on an EAD (EAD 130118-00-0603, 2016)[78]. The high capacity of screws is most efficiently used if they are loaded axially[79].

The variety of screws in terms of their geometrical and steel properties is enormous, ranging from fully threaded screws over screws with a partial thread or two threaded parts to many different head, tip and thread shapes. All these features have different functions. A washer head for instance increases the head pull-through capacity considerably whereas a thread with a large pitch allows for fast insertion. Concerning types of steel, most screws are made

---

[78] In the meantime, the EAD was updated, adding threaded rods: EAD 130118-01-0603, 2019

[79] This explains why specific rules for axially loaded screws have been the sole addition in Eurocode 5, i.e. equations to calculate the tensile, withdrawal and head pull-through capacity and the effective number.

of carbon steel and are hardened. These screws are usually galvanised to protect them against corrosion. Also stainless steel screws are widely applied, where, differently to nails (see e.g. Figure 4-3 on the left), the steel properties differ considerably to those of carbon steel screws because the latter are generally hardened[80]. However, apart from austenitic stainless steel also martensitic stainless steel is used where higher steel properties come with the cost of lesser resistance against corrosion. Moreover, most screws have a coating whose purpose is to reduce friction for easier insertion, which gets more important the higher the density of the used wood species.

Ringhofer (2017) gives a comprehensive and clear overview of these manifold screw types and explains thoroughly the effect of production processes and geometric and material choices on the screw performance.

5.2 Database

The assembled database contains 27793 individual test results taken from 138 test reports on screws from 34 different companies. Only test results evaluated between 2010 and 2019 were considered. All information given in the reports was recorded; from screw classification in terms of geometry and material to specifications concerning the tests, e.g. wood species, angle between screw axis and grain direction or deviations from standard test set-ups. Other screws than self-tapping timber screws (e.g. screws with a metric thread, lag/coach screws) are not discussed here.

Classification of screws

Modern self-tapping timber screws are designated by their nominal diameter d_{nom}, which corresponds to the outer thread diameter d_o, and their nominal length L_{nom}. Designations for partially threaded screws often contain information on the thread length; e.g. a screw "6x400/100" means that the nominal diameter of the partially threaded screw is 6 mm, the nominal length is 400 mm and the (nominal) length of the thread is 100 mm. Moreover, the head type is usually given. The nominal and the measured geometrical properties shown in Figure 5-1 are given in the database[81]. The thread flank angle as a potentially important geometrical value was not measured as this is no required parameter within certification testing. In total, values for 28 different nominal diameters ranging from 2.5 mm to 14 mm were available, where the majority of screws had nominal diameters of 5 mm (14.3%), 6 mm

[80] It is emphasised here that carbon steel screws with a high degree of hardening are prone to hydrogen embrittlement and stress corrosion cracking.

[81] Not always all information is given in the reports; leading to a blank in the database. For instance, it is not necessary to give the pitch p of the thread in "Initial Type Testing (ITT)" in accordance with EN 14592 and hence, p is missing in all considered ITTs.

(17.3%), 8 mm (20.5%) and 10 mm (11.9%). The nominal length of the tested screws was between 20 mm and 2000 mm.

The nominal diameter d_{nom} is an important parameter, as currently, only this value is considered during design (screws are designated using d_{nom}). Other diameters are generally unknown and, if needed, must be taken from the technical documentation of the used screw. Indeed, the variety in screw geometries between companies and even within one company's screw portfolio is enormous. No general rules exist on relationships between different diameters, which would facilitate the use of other than the nominal diameters. As an example, the relationship between inner and nominal diameter is shown in Figure 5-2 on the left[82]. The relationship is best modelled with a quadratic function, but scatter is persistent. For instance, screws with a nominal diameter of 8 mm had inner diameters ranging between 4.82 mm and 6.51 mm. Considering all screws in the database, the inner diameter was between 54% and 85% of the nominal diameter. Consequently, the nominal diameter d_{nom} is indicated using the subscript "nom" in this chapter, to avoid confusion with any other diameter. In analyses, both nominal (d_{nom}) and "measured nominal" (d_{meas}) diameter can be used, where the latter corresponds to the measured outer (thread) diameter d_o. Figure 5-2 on the right shows both diameters[83].

Due to the diversified and often tailor-made screw layout, screw, head and tip types were defined to which all tested screws were assigned, see Table 5-1. All screws were grouped in ***five screw types***, where 68.3% of all screws in the database were partially threaded screws, 20.0% were fully threaded, 6.6% had two threaded parts, 4.8% had a high-low thread (see Table 5-1; two threads with different d_o/d_i ratio) and only 0.3% were screws for timber-concrete composite applications. Concerning the head types, ***four head types*** were specified; countersunk head (74.8%), cylinder head (10.5%), washer head (8.3%) and a special head to fasten steel plates to timber members (6.5%). The classification of the head types is very general, and more exact data on head shapes is not available (see also Figure 5-51 in section 5.6.3). Finally, ***four tip types*** were defined; normal tip (61.5%), drill tip (13.4%), half tip (0.3%) and the remaining 24.9% had special tips of various shapes. This last type encompasses a huge variety of tip shapes, where, however, also the geometry of the other tip types can be rather different, see examples in Figure 5-3. Further specifications given in the database concern the presence of shank ribs or an indentation close to the screw tip. Both features can be seen in Table 5-1, screw type "high-low thread".

[82] Regulations in EAD: $0.5 \cdot d_{nom} \leq d_i \leq 0.9 \cdot d_{nom}$; in EN 14592: $0.6 \cdot d_{nom} \leq d_i \leq 0.9 \cdot d_{nom}$

[83] Outliers were checked and are correct.

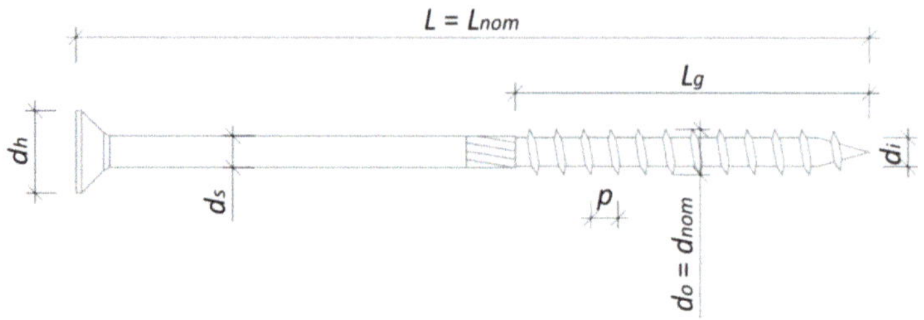

Figure 5-1 Measured and nominal geometrical properties of screws: d_h = head diameter, d_s = diameter of smooth shank, d_o = outer thread diameter = nominal diameter d_{nom}, d_i = inner thread diameter, p = pitch, L = length = nominal length L_{nom}, L_g = length of threaded part.

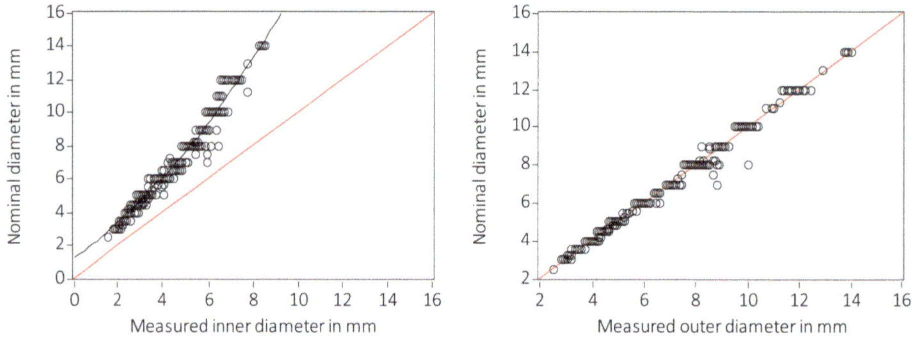

Figure 5-2 Left: Nominal diameter d_{nom} versus measured inner diameter d_i with quadratic trendline in black. Right: Nominal diameter d_{nom} versus measured outer diameter d_{meas}. 27793 screws.

Table 5-1 Classification of screws.

Screw types				
Fully threaded screw Partially threaded screw Two threaded parts Timber-concrete screw (TCC) High-low thread	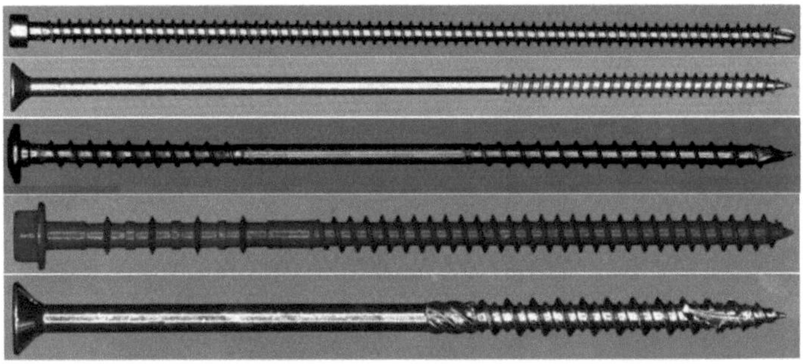			
Head types	Countersunk	Cylinder	Washer	Steel-timber
	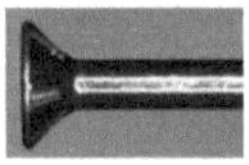		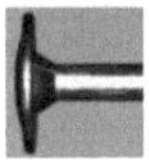	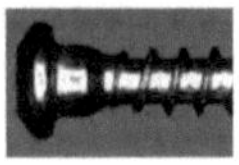
Tip types	Normal tip	Drill tip	Half tip	Special tip
		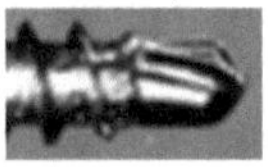	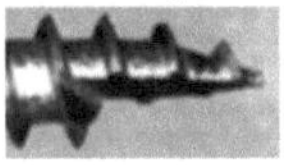	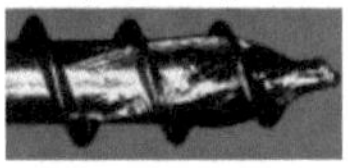

Drill tip

Special tip

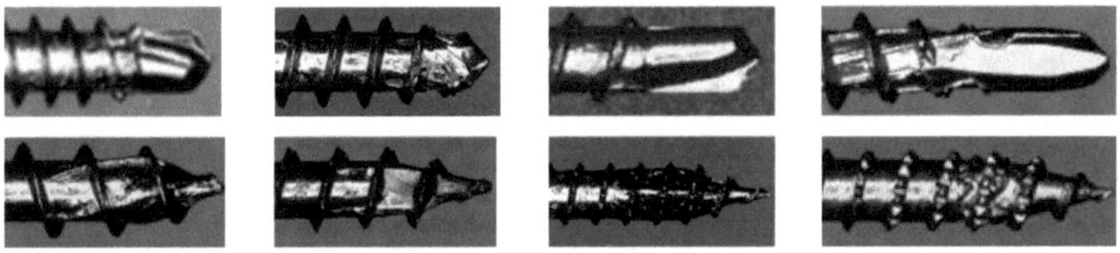

Figure 5-3 Examples of different tip types.

Concerning the types of steel, for 29.9% of all data, the test reports did not explicitly state the types of steel of the used screws, which means, with near-certain probability, that these screws were made from carbon steel, as other types of steel, i.e. stainless steel, are always explicitly mentioned. Therefore, these 29.9% were assigned to carbon steel screws, which then accounted for 83.6% of the database. Generally, it is not explicitly stated if these screws are hardened. In only two reports, it is clearly mentioned that hardened and/or unhardened carbon steel screws were used[84]. It can be safely assumed that most screws were hardened, which is confirmed when considering Figure 5-5. 16.3% were stainless steel screws and only

[84] In one report, steel properties of unhardened screws with nominal diameters of 12 and 14 mm were determined; in the second report, steel properties of screws with nominal diameters of 6, 8 and 10 mm were determined before and after hardening.

40 screws (0.14%) were hot-dip galvanised. No further information about used steel grades was usually given, e.g. which chemical composition was used. In one report, screws before and after hardening were tested. The hardening procedures themselves are never mentioned[85]. Concerning coating, generally, no information was given although the majority of screws is coated. Some reports contained information concerning the type of coating with "Gleitmo 615", a suspension of polymers in water, being most often mentioned. Further in-depth information concerning steel grades, production process, hardening procedures and coatings are given in Ringhofer (2017).

Test results in database

All reported tests were carried out in accordance with the certification rules valid at the time of testing (see also section 2.5). Therefore, apart from the geometric properties, both material and "system" properties were evaluated. Material properties encompass tests on screws (without timber), that determine the screws' yield moment M_y, tensile capacity F_t and torsional moment capacity M_{tor}. System properties are those that include timber, hence insertion moment M_{insert}, withdrawal capacity F_w and head pull-through capacity F_{head}. The used wood species were mainly spruce (*Picea abies*, 65.4%) and laminated veneer lumber made of beech (beech LVL, 20.3%). Other species or types of wood products comprised less than 5% each (OSB, solid wood panel, particleboard, ash, oak, beech, radiata pine, spotted gum, spruce LVL and radiata pine LVL). All timber was stored in a normal climate with 20° C and 65% relative humidity and the moisture content was not measured. This is important to note as the moisture content of beech LVL usually is only around 6% to 8%, whereas that of spruce lies between 10% and 12%. The recorded density was measured as global value of one timber piece, on which more than one test was carried out (see footnote 30). This means that the recorded value must be considered a gross density without any further information concerning the wood characteristics directly around the screw. Mostly, timber was not predrilled (76.5%). Predrilling occurred in 23.5% of all tests and not only hardwoods were predrilled, but also spruce. Predrill diameters ranged between 57% (spruce only) and 85% (beech LVL only) of the nominal diameter. The angle between screw axis and fibre direction was mainly 90° (88.6%) and all head pull-through tests[86] and insertion moment tests[87] were carried out at 90°. Withdrawal tests instead were carried out also at other angles to the grain (from 0° to 90° in steps of 15°). Table 5-2 gives an overview of the tests contained in the database.

---

85 It is important to underline here that martensitic stainless steels can be hardened through a heat treatment, whereas austenitic stainless steels cannot. Austenitic stainless steels may benefit from work hardening, similar to other types of steel, and have a higher corrosion resistance than martensitic steels.

86 Except for 40 tests carried out at 60°.

87 Except for 10 tests carried out at 30°.

Table 5-2 Composition of database. Number in parenthesis is number of series.

	Tensile capacity F_t	Yield moment M_y	Torsional moment capacity M_{tor}	Insertion moment M_{insert}	Withdrawal capacity F_w	Head pull-through capacity F_{head}
Number of tests	3851$^+$ (391)	2921$^+$ (298)	3647 (368)	6863 (524)	7632 (579)	2864 (246)
Of which stainless steel	1085 (114)	756 (81)	1015 (107)	860 (77)	474 (37)	330 (27)
Of which hdg*	10 (1)	10 (1)	10 (1)	10 (1)	–	–
Of which unhardened$^\#$	50 (5)	50 (5)	50 (5)	10 (1)	–	–

$^+$ More values (15 tests) are available, which however are unrealistically high, see also caption to Figure 5-4.

* hdg = hot-dip galvanised.

$^\#$ Only screws made of carbon steel.

5.3 Steel properties

Note: Parts of this chapter were already published in Sandhaas and Blaß (2021).

5.3.1 General

The steel properties to be determined in the framework of certification testing are the yield moment M_y, the tensile capacity F_t and the torsional moment capacity M_{tor}. Generally, 10 tests per property were carried out. The three properties can be analysed individually, which means that individual test results can be considered, e.g. to evaluate if shank ribs have an influence on the tensile capacity of partially threaded screws. Furthermore, steel properties can be compared within test series, e.g. to investigate the relationship between M_y and M_{tor}. For the latter case, only mean values can be used, as there is no direct relationship between individual test values. For instance, the torsional moment capacity cannot be measured on the very screw that was used to determine the yield moment[88]. This issue was already discussed for nails in sections 4.3 and 4.4, where also here, screws may be from different batches. Per screw type, the thread, type of steel and final conditioning (e.g. hardening process) may be the same, but other parameters such as head type and length may differ. As head type and length, among other parameters, are assumed to have no influence on the steel properties, these must not be determined for each individual screw, but the screws can be grouped instead. For example, M_y may be determined on a partially threaded screw with a length of L_1, whereas M_{tor} is determined on the same screw but with a length of L_2. The third property F_t finally is evaluated using again screws with a length of L_1 but with a

88 For long screws, this does not hold; e.g. if long fully threaded screws are considered, steel properties can be determined on the same screw, albeit on different parts of the thread (M_{tor} always with screw head).

different head type. Therefore, experimental values for M_y, F_t and M_{tor} contained in the database are grouped although screw parameters are different, notably length and head type[89]. For certification purposes indeed, these groups are used to derive characteristic screw properties. In terms of bandwidth and representativeness of the database, this variety may be helpful.

It is important to note that the considered ***M_y-value is the value at a bending angle of 45°***, where the bending angle includes elastic bending and eventual machine slip (for an in-depth discussion on this, please refer to section 5.3.2). Furthermore, stainless steel screws are particular as the weakest section may be the smooth shank and not the thread with its smaller inner diameter (work hardening effects, see also footnote 85). However, the reports do not mention where steel failures occurred and only in one report, it is explicitly stated that the yield moment was evaluated at the smooth shank.

The database contains 298 test series with 2921 individual results for M_y, 391 series with 3851 individual results for F_t and 368 series with 3647 individual results for M_{tor}. In total, 265 series are available where all three properties (M_y, F_t, M_{tor}) were evaluated. For 60 series, only results for M_{tor} and F_t are available. In 19 series, only M_y and F_t were tested and five series contain results for M_y and M_{tor}. No comparisons between steel properties can be made for further 94 series as in those only one property was determined. All test results were excluded where failures due to production or screw design errors occurred, e.g. heads were pulled off during a tensile test because the drive was too large. This is done also when calculating characteristic properties in the framework of certification testing. Moreover, unrealistic high values were deleted (see also Table 5-2).

As a first step, the influence of some recorded parameters was investigated. As a result, it can be stated that ***shank ribs*** do not have any influence on the steel properties, which is why this parameter was ignored in all further analyses. The ***type of steel*** instead is known to have an influence, with stainless steel screws having lower properties. This general statement can be confirmed with the database as is shown in Figure 5-4 for the properties "yield moment stress σ_{My}" and "tensile strength f_t". The (extrapolated) trendlines for stainless steel lie below those for (hardened) carbon steel, where however the correlation is error-prone for larger diameters of stainless steel screws due to the lack of test results. σ_{My} and f_t were not measured (M_y and F_t were) and were calculated. Here, σ_{My} was calculated using Eq. (3-1) and f_t using Eq. (4-3), where in both cases the *measured inner diameter d_i of the screws*[90] was used. Moreover, strength values tend to decrease with larger diameters. This was already observed for nails (see Figure 4-2), for which it can be explained with work hardening effects due to cold drawing. Such an explanation does not hold for screws, at least not solely.

[89] Information concerning different parameters was not deleted; screws can still be distinguished.

[90] Only in one report, it is explicitly stated that M_y was evaluated at the smooth shank and hence, the diameter of the smooth shank was considered; f_t and f_{tor} were still evaluated using the inner diameter.

Also for screws, work hardening effects will take place when rolling the thread and consequently, screws with thinner inner diameters (and hence smaller nominal diameters) will "benefit" more from work hardening. Subsequent hardening of screws, a process that includes a heat treatment, may reverse the effect of work hardening, but it will lead to higher properties. This heat treatment may again affect screws with thinner inner diameters more than thicker screws, leading to higher properties of screws with decreasing nominal diameters. In other words, the steel strength is not homogeneous over the cross-section. This can also be the aim of certain hardening procedures such as carbonitriding, where only the outer fibres are hardened to provide a better torsional resistance (Ringhofer, 2017). Also the screw length may influence the steel properties, as the rolling of a long thread may lead to more notches, which in turn lead to reduced properties. However, concerning such a dependency, only a very slight trend is observed in the database, because not enough data with variable screw lengths per diameter is available, and the use of longer screws is correlated with larger diameters.

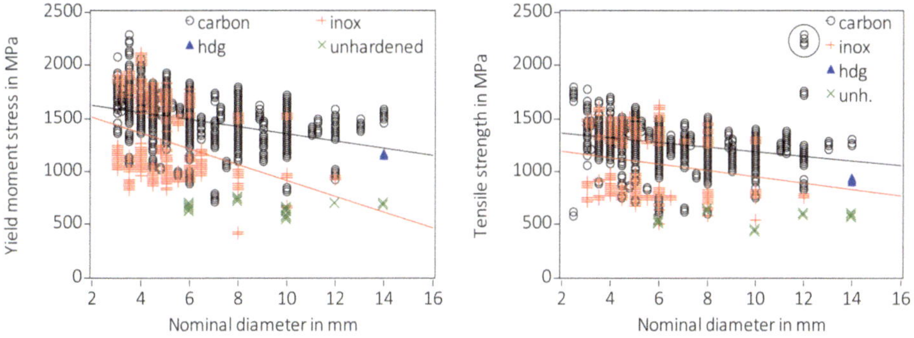

Figure 5-4 Left: Yield strength σ_{My}. Right: Tensile strength f_t (The encircled high tensile strength values of > 2000 MPa for 12 mm screws were deleted, see also Table 5-2), unh. = unhardened. Inner diameters were used to calculate strength values. Trendlines for carbon and stainless steel screws are shown.

Figure 5-4 shows further points. Although the trendline for stainless steel screws lies below that for (hardened) carbon steel screws, stainless steel screws may reach high strength values; i.e. not all stainless steel screws have per se lower mechanical properties than carbon steel screws[91]. As stated in section 5.1, martensitic stainless steels with higher properties are also used for timber screws, and these may reach strength values that are comparable to those of carbon steel screws. This is confirmed when looking at the histograms given in Figure 5-5, where two distinct groups can be identified when looking at stainless steel

[91] The trendline for stainless steel screws in Figure 5-4 on the left may be biased by the outliers at d_{nom} = 8 mm. These low values around $\sigma_{My,mean}$ = 416 MPa stem all from the same report and were calculated using the diameter of the smooth shank, as M_y was evaluated at the smooth shank.

screws. In Figure 5-5, the strength values were calculated from the experimental values of M_y, F_t and M_{tor} and using the *measured inner diameter* d_i: (top) σ_{My} using Eq. (3-1) with d_i, (centre) f_t using Eq. (4-3) with d_i and (bottom) f_{tor} as follows:

$$f_{tor} = \sqrt{3} \cdot \frac{M_{tor}}{Z_{tor,pl}} = \sqrt{3} \cdot \frac{M_{tor}}{\pi \cdot d_i^3 / 12} \tag{5-1}$$

where

f_{tor}	"Torsional strength" in MPa
$\sqrt{3}$	Correction factor to account for difference in shear and tensile strength: $\tau = f_y / \sqrt{3}$. Without correction, strength values cannot be compared.
M_{tor}	Torsional moment capacity in Nmm
$Z_{tor,pl}$	Full plastic polar section modulus of a circular section in mm^3

The "yield moment stress σ_{My}" (top) and the "torsional strength f_{tor}" (bottom) are comparable, whereas both are higher than the "tensile strength f_t" (centre). Different stress states during the different tests may lead to this. For instance, during a test to determine M_y and M_{tor}, the outer fibres are first stressed whereas during a tensile test, the whole cross-section is stressed. As the hardening procedure is not influencing the whole cross-section evenly, the outer fibres with higher strength may lead to higher σ_{My}- and f_{tor}-values. Such a hypothesis should lead to more prominent differences for large diameters, as the hardening should be more efficient for small diameter screws. This can be confirmed when analysing the database; small diameters ($d_{nom} \leq 4$ mm) have slightly higher mean strength values than large diameters[92]. This dependency on the diameter is one of the factors influencing the scatter observed in the strength values shown in Figure 5-5. Strength values for hot dip galvanised and unhardened carbon steel screws are excluded in Figure 5-5; where unhardened screws had f_t-values between 455 MPa and 648 MPa, σ_y = 570...764 MPa and f_{tor} =595...780 MPa. Figure 5-5 shows furthermore that the scatter of steel properties within the whole population of screws is significant. This is in stark contrast to the scatter within the single test series determining M_y, F_t and M_{tor}, where coefficients of variation larger than 0.05 were observed only in 3.5% of the series (see also Figure 5-12 on the left).

[92] Mean strength values in MPa, COV in parenthesis (high for inox due to differences austenitic/martensitic)

	Carbon steel			Stainless steel		
	σ_{My}	f_t	σ_{tor}	σ_{My}	f_t	σ_{tor}
Small d_{nom}	1629 (11%)	1348 (11%)	1647 (14%)	1413 (27%)	1094 (28%)	1357 (30%)
Large d_{nom}	1421 (17%)	1222 (16%)	1447 (17%)	1212 (29%)	1079 (29%)	1286 (27%)

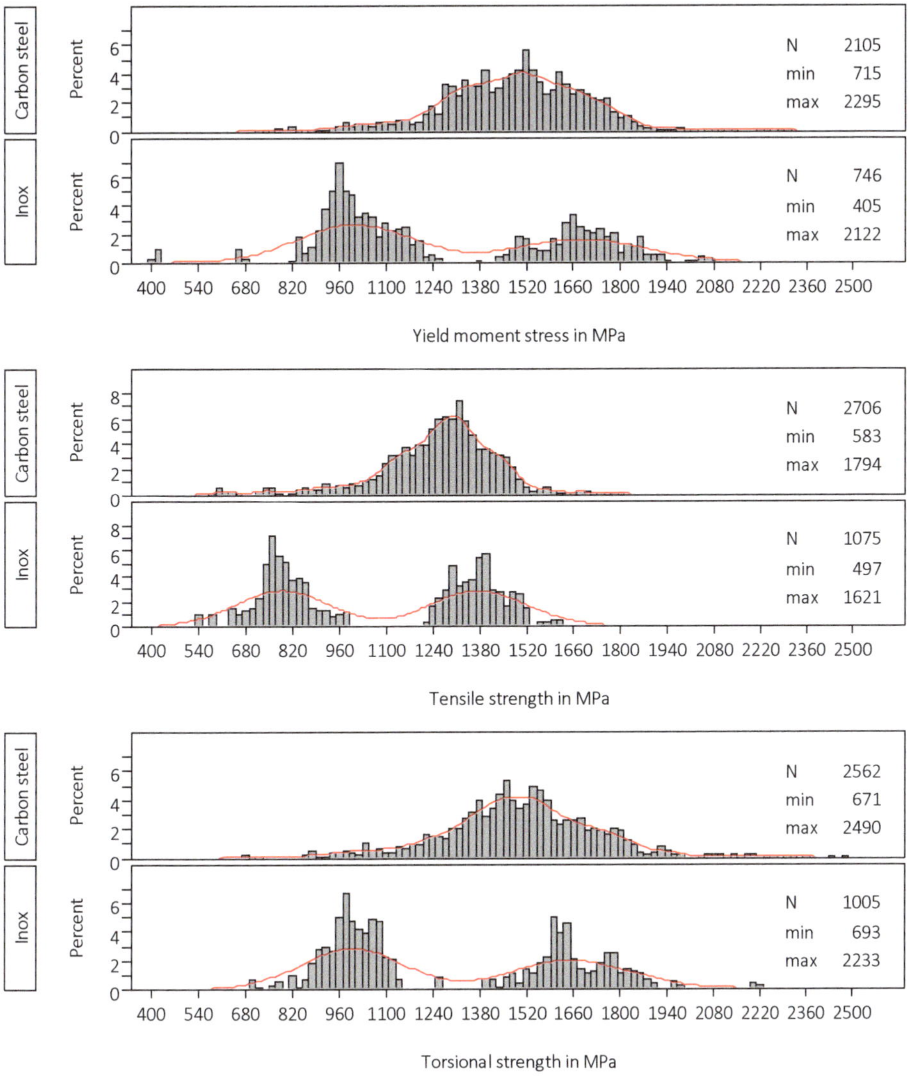

Figure 5-5 Histograms of strength values per type of steel. Results for hdg and unhardened screws are not shown and only results for screws with recorded inner diameter d_i are considered.

Moreover, Figure 5-5 on top shows implications around the already discussed weakest section of stainless steel screws, which may be the smooth shank although its diameter is larger than the inner diameter of the thread. The outliers around $\sigma_{My} \approx 420$ MPa visible in Figure 5-4 on the left are again visible in Figure 5-5 on the top. They all stem from one test report where it is stated that M_y was evaluated at the smooth shank and where hence, the diameter of the smooth shank was considered to calculate the yield strength instead of the (smaller) inner diameter, i.e. leading to lower σ_{My}-values. Also the second group of low σ_{My}-

values around 665 MPa stem all from one report, where however no information is given concerning the location of the M_y-tests and hence, the inner diameter was used to calculate strength values. The same tests led to the lowest values for f_t for stainless steel screws. As a consequence, it is hard to judge the reliability of the upper and lower boundaries of the strength values for stainless steel screws shown in Figure 5-5.

In general, M_y at a bending angle of 45° measured by the testing machine or at rupture is considered in the database (see also sections 2.5 and 5.3.2), whereas F_{tens} and M_{tor} are maximum values, determined without deformation measurements. This statement leads to the question if it is correct to compare the values given in Figure 5-5 or if other M_y-values should be considered; i.e. the value modified in accordance with Blaß et al. (2000), Steilner and Blaß (2014) or at a bending angle of $\alpha = 45/d^{0.7}$. As any comparison done here is based on capacities at failure, i.e. at full plastic capacity, it can be postulated however that it is correct to consider M_y at 45° and not any modified lower value[93].

Apart from the screw diameter, also the ***screw type*** specified in Table 5-1 may influence the steel properties. The following observations made during testing are the reason behind this hypothesis:

- Tensile tests: Screws usually fail in the threaded part with the smallest stressed area (inner diameter). Austenitic stainless steel screws however may also fail in the smooth shank although the diameter of shank is greater than inner diameter. This leads to smaller tensile capacities of partially threaded screws in comparison to fully threaded screws (of the same group). An explanation is that work hardening effects do not occur in the smooth shank of austenitic stainless steel screws.
- Yield moment tests: The weakest section of partially threaded screws usually is in the area of the last thread directly adjacent to the smooth shank (the "transition" area). This may be due to local stress concentrations. Within the same group, partially threaded screws may hence have lower yield moments in comparison to fully threaded screws.
- Often, fully threaded screws have higher steel properties than partially threaded screws, as the latter are less hardened[94].

Figure 5-6 shows all three steel properties versus the nominal diameter (excluding data for hdg and unhardened screws as those lead to a bias in the regressions). Nearly 90% of all screw types were partially threaded and fully threaded screws. The black and red lines in Figure 5-6 are the quadratic trendlines for partially threaded screws (the red line) and for fully threaded screws (the black line). Considering that the trendlines for nominal diameters

93 Comparisons of steel properties are still not fully correct, as hardening effects are not considered. Furthermore, some screws can reach even higher M_y-values for bending angles > 45°, although in general, moment-rotation curves are only weakly inclined once plastic behaviour is reached, see also Figure 2-3.

94 Fully threaded screws need high steel properties to allow for insertion without breaking.

larger than 12 mm are not reliable (few test results and no results for partially threaded screws), the two lines do not differ significantly. Therefore, on the level of individual test results, no difference between screw types can be found. Screws with a nominal diameter of 8 mm, the dominating diameter in the database, however seem to form distinct groups, where partially threaded screws scatter less and do not reach low values.

Figure 5-6 on the top shows additionally the mechanics-based plastic bending moment M_{mech} calculated in accordance with Eq. (5-2).

$$M_{mech} = \frac{1}{6} \cdot 1200 \text{ MPa} \cdot (0.64 \cdot d_{nom})^3 \quad (5\text{-}2)$$

where

1200 MPa Mean value of tensile strength of all screws (mean of 3846 individual values, COV = 21%), calculated using the inner diameter d_i: $f_{t,di} = F_t \cdot 4 / (\pi \cdot d_i^2)$

0.64 0.64 is the mean ratio of d_i/d_{nom} (mean of 27282 individual values, COV = 5.6%)

The current state of knowledge postulates that the yield moment of self-tapping screws must be experimentally determined, as it cannot be calculated using standard equations such as Eq. (4-1). The complex screw geometry, work hardening effects through cold forming and hardening processes lead to complex stress states in the screw cross-section when it is bent and hence the yield moment may not be easily captured by simple models. Moreover and as discussed in depth in Ringhofer (2017), the stressed area to consider when transforming capacity into strength is not even circular in the case of screws. However, Eq. (5-2) was applied to verify the predictive quality of such a simple equation, and it corresponds astonishingly well to the trendlines considering that one mean value for tensile strength and "inner diameter" was used. This result motivates a closer look into possible simple calculation models valid for screws. Such an analysis could be made in analogy to section 4.4. The potential result would be a regression equation allowing for the calculation of characteristic values, which is based on tensile strength and diameter as input. Such an approach could furthermore allow for elimination of yield moment tests, which, considering difficulties with testing (see next section 5.3.2), would eliminate an uncertain parameter.

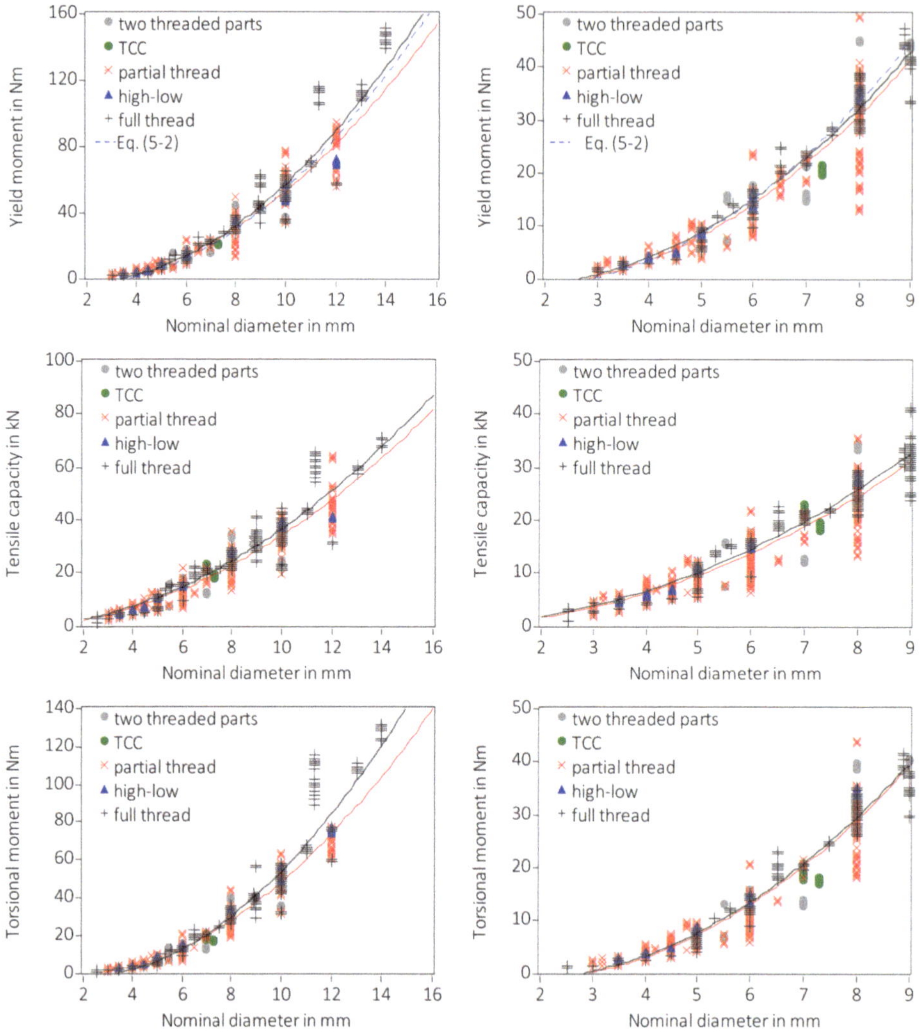

Figure 5-6 Properties versus nominal diameter d_{nom}, identified by screw type and excluding data for hdg and unhardened screws. From top to bottom: Yield moment M_y (69.3% partial thread, 17.5% full thread), tensile capacity F_t (70.4% partial thread, 17.4% full thread), torsional moment capacity M_{tor} (66.6% partial thread, 20.5% full thread). The black and red lines are the quadratic trendlines for partially threaded (red) and fully threaded (black) screws. The left and right columns differ in scale.

5.3.2 Discussion of test procedures, using the example of M_y

Before further analysing the steel properties, issues arising from test specifications must be addressed. Lack of clarity in testing standards is indeed hampering proper analysis of test data. Although the tests contained in the database stem all from the same laboratory and hence, tests were carried out using the "same" methodology and equipment, changes in testing standards lead to difficulties when comparing results. This issue was already addressed in section 4.5, where changing requirements concerning the selection of timber species for withdrawal tests were discussed. Concerning screws, the determination of the yield moment changed over the years. However, the prescribed test method was always a four-point-bending test. In the first assembled testing reports, the yield moment was determined at the maximum bending angle ≤ 45° measured by the testing machine. In order to obtain characteristic values, this raw testing data must then be modified. In the past, the value was modified applying the equation given in Blaß et al. (2000). Later, the method proposed by Steilner and Blaß (2014) was applied to change the shape of the moment-rotation curve in order to consider only plastic bending. Meanwhile, the EAD for screws (EAD 130118-00-0603, 2016) states that the yield moment "is the value at the ***plastic*** bending angle $\alpha = 45/d^{0.7}$ degrees"[95], whereas neither EN 409 (2009) nor EN 14592 (2012) mention the word "plastic". Still, both EAD and EN 14592 require a reduction of the angle α at which the yield moment must be read; i.e. $\alpha = 45/d^{0.7}$. Moreover, EN 409 prescribes different bending angles for different fastener types and timber products, leading to different bending angles for different wood densities. All these changes in the course of time make comparisons difficult. This explains, why raw data measured by the testing machine at a bending angle of 45°[96] is recorded in the database. Hence, any calculated characteristic values will be different from the values given in technical documentation unless the raw data is corrected as explained above.

Looking at the broader picture of potential future databases containing results of different laboratories, which can be used for stochastic analyses, vague testing standards exacerbate the problem. For instance, in EN 409 (2009), the test location along the screw axis is not specified. Especially for partially threaded screws, the exact definition of this location is necessary in order to determine M_y in the weakest section. Furthermore, different test setups are conceivable when consulting EN 409, which prescribes a four-point bending test. Above all, it is not clearly stated that the screw must be allowed to move horizontally, so that there is scope to find the weakest section. Additionally, the range of the free length L_2 between the two load insertion points, which influences M_y-values, is with d and $3 \cdot d$ too large (see also Figure 5-8). Another issue not addressed by the testing standard is the fact that most test setups measure the global bending angle, whereas only the plastic (not elastic) bending angle should be considered. Steilner and Blaß (2014) addressed this issue and proposed a

95 One of the reasons for reducing M_y is the assumption that the full plastic bending capacity is not reached in a real joint as the fasteners will not deform sufficiently (see also Blaß et al., 2000).

96 In the database, the M_y-values given were measured values at rupture or at a global maximum bending angle or at 45° (including elastic and plastic deformation), determined using the moment-rotation curves of the testing machine and hence including further errors due to e.g. machine slip. The method proposed by Steilner and Blass (2014) was only used for the most recent 11 testing reports with in total 30 test series for M_y. For these series, the original moment-rotation curves of the testing machine are not available.

solution to determine the plastic bending angle. However, machine slip is still included, as the standard test setup used at KIT measures the bending angle directly through the rotation of the machine[97]. This is less of an issue because registered values for M_y are generally very close to the plateau value (full plasticisation), where only a very slight increase in yield moment with increasing bending angle is observed (see also Figure 2-3 on the left).

Different laboratories may therefore determine different M_y-values for the same screws. To look into this, some tests in accordance with EN 409 were carried out in two different laboratories. The location, where M_y was to be tested, was determined beforehand. The few preliminary tests presented here must be considered as being illustrative; no thorough test planning nor a properly documented test execution was taking place. Figure 5-7 shows these few tests using seven different screws and one steel dowel. Concerning the location, fully threaded screws and dowels were tested in the centre; partially threaded screws were tested "at the transition" between thread and smooth shank. One partially threaded screw was tested at three locations: in the centre of the thread and of the smooth shank and at the transition. It can be seen that, although the exact test location was defined, one laboratory obtained higher values for partially threaded screws tested at the transition, with a maximum difference of 35%. The laboratory obtaining the smaller values tested "in the last thread directly adjacent to the transition" as is shown in Figure 5-7; the other laboratory tested "close to the shank ribs of the transition", obtaining in general higher values for M_y.

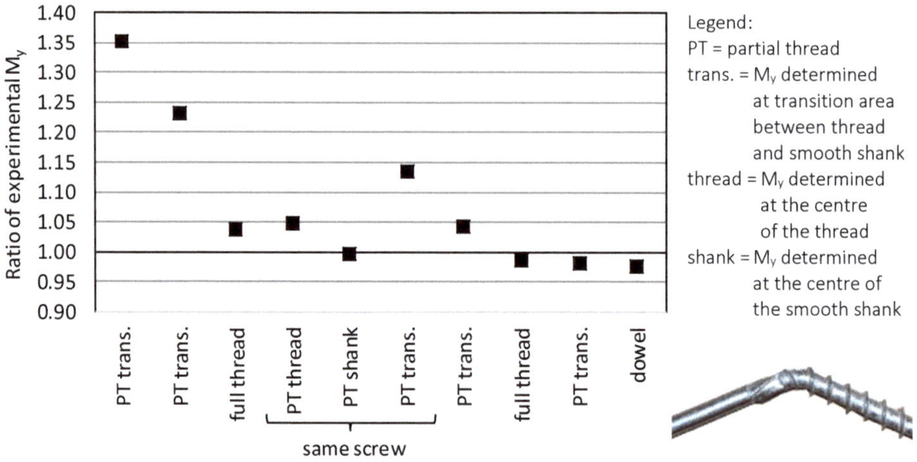

Figure 5-7 Yield moments determined in two laboratories. Exemplary photo of weakest section of partially threaded screws ("last thread directly adjacent to the transition").

[97] This slip is reduced as much as possible and a calibration was carried out to eliminate machine slip in the experimental curves.

This discussion can be underlined with test results presented in Kuck and Sandhaas (2022), where bending tests were carried out at different screw cross-sections and with different free length values L_2. The 160 mm long screws had a nominal diameter of d_{nom} = 8 mm and a partial thread with a thread length of L_g = 80 mm (d_s = 5.8 mm, d_i = 5.4 mm). The results are given in Figure 5-8, where the procedure in accordance with Steilner and Blaß (2014) was applied to subtract the elastic bending deformation from the global moment-rotation curve. Concerning different screw cross-sections, obviously, the yield moment is higher in the smooth shank section with its larger diameter. The general shape of the M-α-curve shows a nearly perfectly plastic behaviour for the results in the transition area and in general stiffer results for the smooth shank area. Concerning the free bending length, the tests with "no free bending length L_2 = 0"[98] and hence small bending radii result in higher moments at smaller bending angles, as then the full plasticisation of the cross-section is reached more swiftly. The different bending radii can be seen in Figure 5-9 where four screws after a bending test are shown together with a screw taken from the birch joint shown Figure 2-2. More simple three-point bending tests, e.g. in accordance to ASTM F1575/F1575M (2021) or as presented in Bader et al. (2016) could be a viable option to overcome at least some of the testing issues of four-point bending tests in accordance to EN 409 (2009).

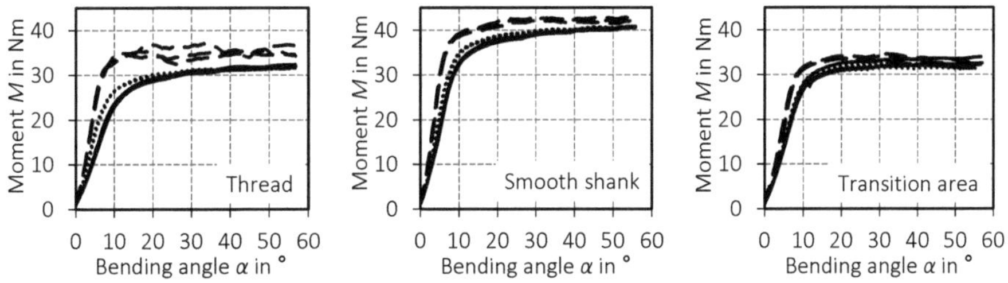

Figure 5-8 Moment-rotation curves for partially threaded screws with d_{nom} = 8 mm determined at different locations "thread", "smooth shank" and at the "transition area" and with varying free bending length L_2: —— L_2 = 30 mm; ······ L_2 = 2 d_{nom}; – – L_2 = 0. The curves were adjusted for their elastic component in accordance with Steilner and Blaß (2014). (Kuck and Sandhaas, 2022)

[98] Obviously, L_2 = 0 is not possible, it was chosen to be as small as possible. Likewise, the largest free bending length L_2 = 30 mm used during the tests was the largest L_2 that could be accommodated by the test rig.

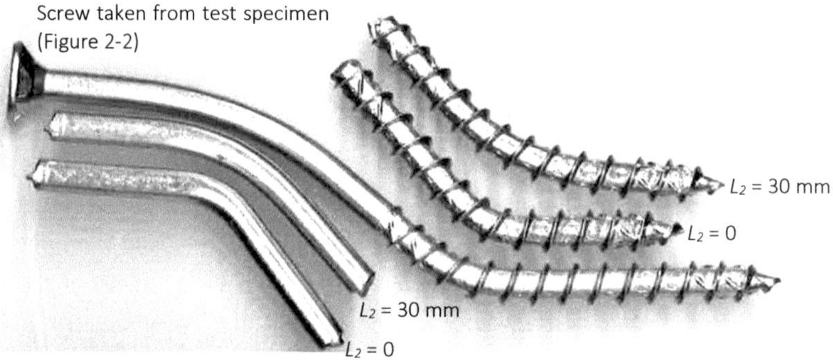

Figure 5-9 Comparison of bending radii at L_2 = 0 and L_2 = 30 mm and screw taken from joint test specimen made of birch glulam after failure (Figure 2-2). (Kuck and Sandhaas, 2022)

5.3.3 Comparison between M_y and M_{tor}

Considering these difficulties in determining an experimental value for M_y, the best would be to abandon the tests and, if possible, to determine the yield moment indirectly. Recalling section 4.4, where the yield moment of nails was analysed, an analogous analysis could be carried out here, where the yield moment of screws is considered being a function of the tensile strength and the diameter. For screws, however, this is not as straightforward as for nails, because the definition of "the diameter" and the steel properties of the cross-section are more complex. For instance, the difference between inner and outer (nominal) diameter is much larger for screws, with the inner diameter being between 54% and 85% of the nominal diameter. Moreover, the diameter is not only needed explicitly in any possible equation, but also implicitly, as the experimentally determined tensile capacity must be transformed into a tensile strength. Finally, the efficacy of hardening procedures of carbon steel screws is depending on the screw diameter, with small diameter screws having more evenly distributed steel properties over their cross-section than large diameter screws, where the outer part of the cross-section may show higher strength than the inner part[99].

Another idea is given by Figure 5-5, which shows that the stress data calculated from the yield moment is similar to the strength data calculated from the torsional moment capacity[100]. A straightforward approach without any need of geometrical data would hence be to link the yield moment M_y directly with the torsional moment capacity M_{tor}. Tests to determine the torsional moment capacity are much easier to execute than tests to determine the yield moment, and inaccuracies as discussed in section 5.3.2 are less likely to occur. Above all, the torsional moment capacity is determined *independently of any deformation*; it is the maximum moment measured before the screw breaks. The torsional moment capacity is not

99 Here, the discussion in section 5.3.1 is recalled, including Figure 5-5 and footnote 92.

100 Rainer Görlacher was the first to propose this.

needed for design; the test is carried out to make sure that screws can be drilled in without breaking. For this purpose, M_{tor} is compared with the insertion moment M_{insert} (section 5.4).

If M_{tor} is directly compared to M_y, the torsional moment capacity tends to be slightly lower as can be seen in Figure 5-10 on the left. In total, 270 series were considered, as only for these series both M_y (2646 individual values) and M_{tor} (2670 individual values) were determined. Of the 270 series, 241 series (89%) had higher M_y-values than M_{tor}-values. However, a direct comparison between yield moment and torsional moment capacity cannot be done from a mechanical point of view, as a four-point-bending test leads to normal stresses in the screw, and a torsional test to shear stresses (see also correction factor in Eq. (5-1)). Furthermore, not only the "stress type" is different, but also the section moduli differ. As a consequence, a corrected $M_{tor,corr}$ was used in Figure 5-10 on the right:

$$M_{tor,corr} = \sqrt{3} \cdot M_{tor} \cdot \frac{Z_{pl}}{Z_{tor,pl}} = \sqrt{3} \cdot M_{tor} \cdot \frac{d^3/6}{\pi \cdot d^3/12} = \sqrt{3} \cdot M_{tor} \cdot \frac{2}{\pi} \approx 1.1 \cdot M_{tor} \quad (5\text{-}3)$$

where

$M_{tor,corr}$ Corrected torsional moment capacity in Nmm

$\sqrt{3}$ Correction factor to account for difference in shear and tensile strength: $\tau = f_y/\sqrt{3}$.

M_{tor} Experimental value for torsional moment capacity in Nmm

Z_{pl} Full plastic section modulus of a circular section in mm^3

$Z_{tor,pl}$ Full plastic polar section modulus of a circular section in mm^3

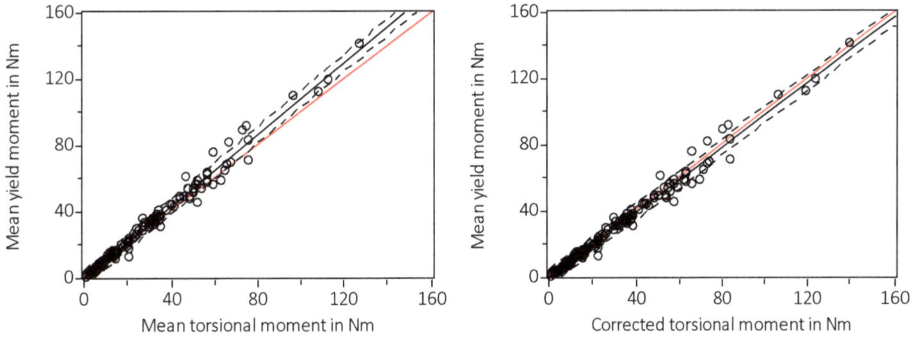

Figure 5-10 Left: Yield moment versus torsional moment capacity. Right: Yield moment versus corrected torsional moment capacity. Mean values of 270 test series. Red line is bisect line, black line is trendline, dotted line is 95% confidence interval.

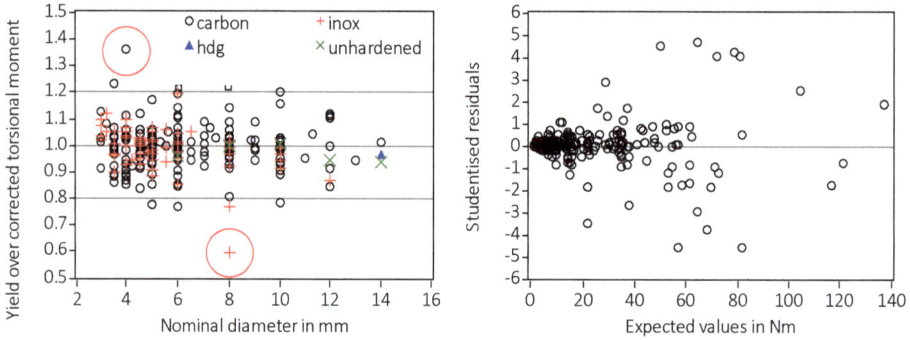

Figure 5-11 Left: Yield moment divided by corrected torsional moment capacity. Right: Studentised residuals versus expected values of Eq. (5-4)

Apparently, the accordance between M_y and $M_{tor,corr}$ is better than between the uncorrected mean values. In Figure 5-11 on the left, the data of Figure 5-10 on the right is again shown, but in terms of M_y divided by $M_{tor,corr}$ versus nominal diameter. The mean value of all 270 ratios is 0.99 with a coefficient of variation of 8.3%. The two outliers encircled in red were checked. The upper circle identifying the highest ratio of 1.36 belongs to a test series, where the yield moment was determined on screws with a length of 50 mm, and the torsional moment capacity on screws with a length of 70 mm.

The lowest ratio of 0.59 instead stems from a series where the same screws were tested. The screws were stainless steel screws with a mean tensile and yield strength value of 580 MPa and 508 MPa, which are low values considering Figure 5-5, and a considerably higher mean torsional strength of 858 MPa. It is unknown if maybe the same screws, but from a different batch were used.

The scatter of the ratios in Figure 5-11 on the left is however too high to simply apply Eq. (5-3) in order to calculate the yield moment M_y as $1.1 \cdot M_{tor}$. The horizontal reference lines at 1.2 and 0.8 show that consistent differences between mean values of M_y and $M_{tor,corr}$ of ±20% are present. The scatter of the ratios however seems to decrease slightly with larger diameters. Transforming mean values into characteristic values calculated in accordance with EN 14358 (2016) and recalculating the ratio does not change the general picture. Another path to find a relationship between M_y and M_{tor} is a statistical analysis and hence, a linear regression was carried out leading to the following equation (R = 0.99):

$$M_{y,mean} = 1.08 \cdot M_{tor,mean} + e \quad (5\text{-}4)$$

where

$M_{y,mean}$	Expected value of mean yield moment
$M_{tor,mean}$	Experimental mean value for torsional moment capacity (independent variable)
e	Residuals

Eq. (5-4) confirms the theoretical outcomes of Eq. (5-3). A benefit of any regression analysis is that a residual analysis can be carried out, in order to gain a better understanding of the data. Figure 5-11 on the right shows the studentised residuals versus the expected values $M_{y,mean}$, where nine series have studentised residuals larger than |3| and should hence be excluded (Hartung, 2009, p. 586). Looking at the distribution of the studentised residuals however, it gets clear that excluding all values larger than |3| is a never-ending story, as due to the cluster of values at small expected $M_{y,mean}$, further studentised residuals come to lie above |3| with each new iteration step. The distribution indicates that larger expected values, i.e. screws with larger diameters, have larger studentised residuals[101]. This is confirmed when looking at statistical diagnostic plots implemented within SAS that show that only screws with nominal diameters ≥ 10 mm (and three series with d_{nom} = 8 mm) have critically high leverage and Cook's distances[102]. When considering the clear linear trend of Figure 5-10 and the symmetric scatter of the ratio of ±20% in Figure 5-11 on the left, the linear regression model itself is appropriate and hence not the cause for these statistical shortcomings.

When re-examining the database, it becomes clear that for screws with d_{nom} ≥ 10 mm, M_y and M_{tor} were tested on screws with a different length in 12 of 43 series (28%), whereas only for 28 of 227 series (12%) with screws with d_{nom} < 10 mm, different screws were used. This is a (slight) indicator that, similar to nails, properties of screws with a certain length should not be extrapolated to the "same" screws of a different length. Here, the general production process of a screw can be recalled, where long threads are rolled in parts, leading to potentially unsmooth transitions within long thread lengths. Obviously, seeing the large difference of available test data (43 series versus 227 series), there is also a clear lack of testing data for screws with larger diameters, that may influence regression analyses.

Concerning the variation of the test results, no difference can be observed within the different test series. Figure 5-12 on the left shows standard deviations of all series determining M_y and M_{tor}, where 3.4% (M_y) and 3.5% (M_{tor}) had standard deviations higher than 0.05. The standard deviations were calculated in accordance with EN 14358 (2016) and based on a lognormal distribution. Standard deviations seem to decrease with larger diameters, indicating that not enough test results are available. Contrary to nails, one property was usually evaluated using the same screws, i.e. of the same length, same package, same batch. Indeed, for nails, observed standard deviations were higher (see Figure 4-5 on the right), which can be explained with the fact that nails from different batches were tested within one series.

[101] Similar to the residual analysis carried out in section 4.4.1, simple statistical tools that test for heteroscedasticity do not identify this; e.g. the VIF is 1.0.

[102] Of the available 43 series on screws with d_{nom} ≥ 10 mm, 38 had high leverage, were identified as outliers or a combination of both.

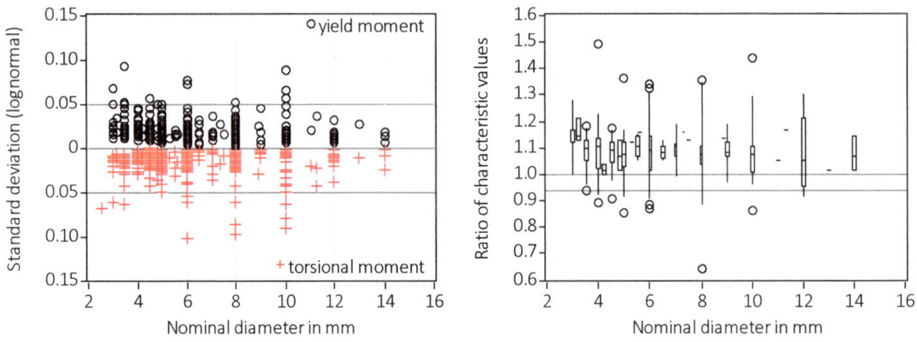

Figure 5-12 Left: Standard deviation s_y of 298 test series for M_y and 368 test series for M_{tor}. Right: Boxplot of ratios of characteristic yield moments divided by characteristic torsional moment capacities, which were calculated in accordance with EN 14358 (2016) using uncorrected M_{tor}-values. 5th and 95th percentiles are specified. Horizontal line at 0.94 indicates observed 5th percentile.

Therefore, complex four-point bending tests could be replaced by much simpler torsional tests, which are independent of any deformation measurements. The motivation for this statement are the comparisons shown in Figure 5-10 and the accordance between theoretical and statistical approach (Eqs. (5-3) and (5-4)). The observed scatter when comparing M_y-with corrected M_{tor}-values (Figure 5-11) is probably due to the following main issues:

- Data for large diameter screws are missing.
- Properties were not always determined on the same screws (i.e. same batch, same length).
- Dependency of M_y-values on deformations and hence more difficult to determine than the ultimate value of M_{tor}.
- Effect of hardening is unknown.

Despite these issues, characteristic values for M_y need to be derived in order to provide for safe design values. As M_y was generally slightly higher than M_{tor}, see Figure 5-10 on the left, this natural safety margin is utilised. This means that uncorrected M_{tor}-values were used to calculate 5th percentiles in accordance with EN 14358 (2016) and based on a lognormal distribution. These $M_{tor,k}$-values then represent predicted $M_{y,k}$-values. Analogously, 5th percentiles for the yield moment were calculated for each of the available 270 series. Figure 5-12 on the right shows a boxplot of the ratio of characteristic values, $M_{y,k}$ divided by $M_{tor,k}$, versus the nominal diameter. The observed 5th percentile is 0.94. This means that too many predicted $M_{y,k}$-values (equal to $M_{tor,k}$) are higher than the experimentally based $M_{y,k}$-values. An additional conversion factor of 0.94 is hence needed. Eq. (5-5) shows a possible equation to calculate $M_{y,k}$ based on (uncorrected) $M_{tor,k}$-values.

$$M_{y,k} = 0.94 \cdot M_{tor,k} \tag{5-5}$$

where

$M_{y,k}$ Characteristic value of yield moment

$M_{tor,k}$ Characteristic value of torsional moment capacity (based on tests)

Simply eliminating tests to determine the yield moment is not constructive however. The ability of screws to bend without rupture cannot be validated with torsional tests. Current commonly accepted design practice asks for sufficient ductility of joints and hence sufficient elongation of screws[103]. Independently of any possible changes in methodology, the findings of this section allow for plausibility checks, as a comparison of M_y- and M_{tor}-values can help to identify erroneous values.

5.3.4 Calculation of yield moment using tensile strength

In analogy to section 4.4.1, nonlinear regressions were carried out using the model approach given in Eq. (4-4), where for nails, the measured nominal diameter was used. All individual steps are analogous to those taken in section 4.4.1, excluding the generation of simulated values. The procedure was iterative, eliminating all outliers with studentised residuals larger than |3|. The regression was carried out based on mean values. Table 5-3 gives results of four different nonlinear regressions, where different screw diameters were used to calculate the tensile strength f_t. Model A used the inner diameter d_i to calculate f_t and to perform the regression. This model is thought to be closest to the true mechanical behaviour, although, as discussed before, screws do not have a circular cross-section with a constant diameter and any stress concentrations are not considered. Model B used the measured nominal diameter d_o (outer diameter), analogously to the regression carried out for nails. Model C is hybrid, because d_i is used to calculate f_t and d_{nom} is used for the regression. The idea behind model C is that it would be possible to give a tensile strength instead of a tensile capacity in technical documentation, which would most probably be calculated with d_i. In design then, the yield moment would be calculated using a known diameter, i.e. the nominal diameter. Model D uses only the nominal diameter of the screws. Finally, model E is similar to model D. It however forces the exponent of the nominal diameter to be equal to 3. It is recorded here that the ratio between inner and nominal diameter was between 0.54 and 0.85. For all five models, the regression results before and after exclusion of outliers are given in order to indicate the influence of statistical outliers, which may not be real outliers but rather correct representations of the true scatter. Model A was furthermore applied using different

[103] There is no agreement on what "sufficient ductility" is, see also section 2.2. Currently, the (EAD 130118-00-0603, 2016) for screws asks for 10 mm joint displacement before failure, if reduced spacing rules for joints with laterally loaded screws are requested.

subgroups (only carbon steel, austentic or martensitic stainless steel screws), which led to slightly different regression variables (max $\Delta\alpha$ = +0.047 and max $\Delta\beta$ = -0.1024, regression included outliers) whilst still having R^2-values of 0.99.

Table 5-3 Results of nonlinear regression for screws in accordance with Eq. (4-4). The bold model results are shown in Figure 5-13.

Model		No of series	Regression variables α	β	R^2	Excluded series
A	$M_{y,mean} = \alpha \cdot f_{t,di,mean} \cdot d_i^{\beta}$	256	**0.1875**	**3.0192**	**0.999**	27
		283	0.1771	3.0521	0.993	-
B	$M_{y,mean} = \alpha \cdot f_{t,do,mean} \cdot d_o^{\beta}$	241	0.1519	2.9006	0.998	42
		283	0.1526	2.9052	0.988	-
C	$M_{y,mean} = \alpha \cdot f_{t,di,mean} \cdot d_{nom}^{\beta}$	261	0.0884	2.7125	0.991	22
		283	0.0977	2.6748	0.963	-
D	$M_{y,mean} = \alpha \cdot f_{t,dnom,mean} \cdot d_{nom}^{\beta}$	239	**0.1486**	**2.9105**	**0.999**	44
		283	0.1493	2.9101	0.989	-
E	$M_{y,mean} = \alpha \cdot f_{t,dnom,mean} \cdot d_{nom}^{3}$	238	0.1234	-	0.998	45
		283	0.1208	-	0.988	-

Figure 5-13 on the left shows the experimental and the expected values for model A (with outliers excluded). The results are presented as the ratio of experimental over expected values in order to better identify differences between both values, seeing that the high R^2 = 0.999 would not allow to see these differences if the values are plotted versus each other. Data for higher nominal diameters are scarce and the ratio scatters more for smaller diameters[104]. In general, experimental values are rather underestimated, as the ratios tend to be above 1.0. The mean value of the ratio is 1.04 and the observed 5th percentile is 0.94 with a COV = 6.7%. Maximum differences of 28% were observed. Figure 5-13 on the right shows again the experimental results differentiated by the type of steel and the curves for three models, where the models are based on mean values and should hence capture the mean experimental results. Model A is again shown, but now in terms of calculated values in accordance with Eq. (4-4) using the variables α and β given in Table 5-3, with $d_i = 0.64 \cdot d_{nom}$ and $f_{t,di,mean}$ = 1200 MPa (see Eq. (5-2)). For model D, the nominal diameter was considered in Eq. (4-4), and the mean tensile strength was calculated using the nominal

[104] Of all 256 series, the most frequently used nominal diameters were 5 mm and 6 mm with 37 series each, 32 series were with 4 mm, 4.5 mm and 8 mm were tested in 31 series each, 22 series were with 3.5 mm and 17 with 10 mm.

diameter, which lead to a value of $f_{t,dnom,mean}$ = 484 MPa (COV = 22%, 3856 individual values). Finally, also the curve in accordance with the current equation given in Eurocode 5, Eq. (4-1), is given, where all input values are referred to the nominal diameter (as in model D). Figure 5-13 on the right shows that the regression results do not differ much, which implies that the nominal instead of the inner diameters can be used in such equations. Eq. (4-1) is less steep, hence works less for smaller diameter screws and better for larger diameter screws. However, all of the models overestimate the yield moments for larger diameter screws, at least considering the few available test results. This is, probably, a consequence of decreasing tensile strengths with increasing diameters as shown in Figure 5-4 on the right.

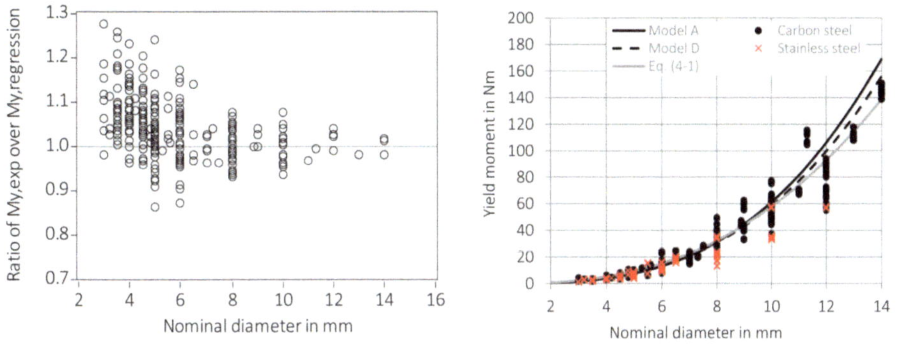

Figure 5-13 Left: Ratio of experimental $M_{y,mean}$ over expected value of model A excluding outliers. Right: Individual experimental results (2921 values, hdg and unhardened excluded) and calculation results using the regression variables given in Table 5-3: Model A with $d_i = 0.64 \cdot d_{nom}$ and $f_{t,di,mean}$ = 1200 MPa, Model D with d_{nom} and $f_{t,dnom,mean}$ = 484 MPa, current model from EC 5 = Eq. (4-1) with d_{nom} and $f_{t,dnom,mean}$.

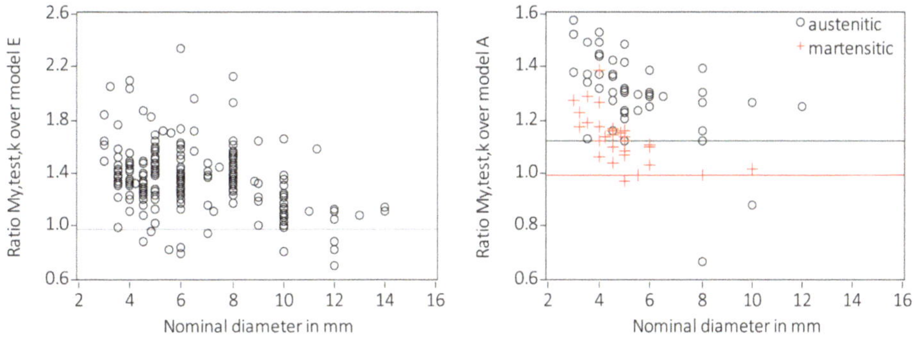

Figure 5-14 Left: Ratio of characteristic values over $M_{y,k} = 0.1234 \cdot 365\ \text{MPa} \cdot d_{nom}^3$. Only test results for hardened screws made of carbon steel are considered. The horizontal line at 0.98 indicates the observed 5th percentile. Right: Ratio of characteristic values over model A. Group "austenitic stainless steel screws" with $f_{t,di,k}$ = 633 MPa: $M_{y,k} = 0.1875 \cdot 633\ \text{MPa} \cdot d_i^{3.0192}$. Group "martensitic stainless steel screws" with $f_{t,di,k}$ = 1254 MPa: $M_{y,k} = 0.1875 \cdot 1254\ \text{MPa} \cdot d_i^{3.0192}$. Horizontal lines show observed 5th percentiles.

If the aim of any possible design rule is to make consultation of Declarations of Performance superfluous, then a general characteristic value for the tensile strength valid for all screws of a certain type of steel is needed. For instance, 365 MPa, which is the observed 5th percentile of the tensile strength $f_{t,dnom,k}$ of hardened screws made of carbon steel can be chosen as characteristic value. Now, model E is chosen to check regression results, and $f_{t,dnom,k}$ = 365 MPa is inserted instead of $f_{t,dnom,mean}$. The result is shown in Figure 5-14 on the left, where it is compared with characteristic values determined in accordance with EN 14358 (2016) and where only hardened screws made of carbon steel were considered. The mean value of the shown ratio is 1.35 (COV = 19%) and the observed 5th percentile is 0.98. Therefore, the chosen value for $f_{t,dnom,k}$ = 365 MPa allows for a calculation of $M_{y,k}$. Figure 5-14 on the right shows analogous results, but considering model A and two groups of stainless steel screws; "austenitic" with $f_{t,di} < 1100$ MPa and "martensitic" with $f_{t,di} > 1100$ MPa (see Figure 5-5 in the centre). The observed 5th percentiles of the tensile strength were $f_{t,di,k}$ = 633 MPa (540 observations, COV = 11%) for group "austenitic" and $f_{t,di,k}$ = 1254 MPa (535 observations, COV = 6%) for group "martensitic". Whereas model A works well for the group "martensitic" with an observed 5th percentile of the ratio of 0.99, its predictive capability is less good for the group "austenitic" with a 5th percentile of 1.12. Persistent large scatter can be observed and the accumulation of data for nominal diameters ≤ 6 mm. Figure 5-14 emphasises the challenges applying such a conservative approach using a single strength value, as the large variety of screws cannot be represented.

5.3.5 Conclusions regarding the steel properties

The test results contained in the database showed clearly the challenges concerning steel properties of screws as a result of their variety. Contrarily to staples and nails, self-tapping screws are much more versatile in terms of used steel grades and productions methods, and also hardening procedures impact considerably on their diversity. Therefore, tensile and torsional tests will remain necessary to establish these two basic properties. Concerning the yield moment, design equations of manifold shapes are possible, but all of them can reproduce only a conservative value for $M_{y,k}$, if the whole population of self-tapping screws is considered. The observed coefficient of variation within the single test series is approx. 5% as can be expected when dealing with steel properties, see also Figure 5-12 on the left. Between test series however, for instance considering screws with a nominal diameter of 8 mm, observed coefficients of variation were approx. 12% for carbon steel screws and approx. 28% for stainless steel screws. Consequently, the derivation of general equations valid for the whole screw population comes with the cost of conservative values for certain screws leading to wrong predictions of joint failure modes, which will be problematic e.g. in seismic design. Particularly screws made of stainless steel should, at least, be divided in austenitic and martensitic stainless steel. If such a simple approach is wished for and if no differentiation between diameters is made, the database must be extended with data for larger diameters, as any design equation derived does not hold for diameters larger than ca. 10 - 12 mm, which gets particularly important if exponential approaches are chosen.

Concerning the experimental determination of the yield moment, EN 409 (2009) should be reviewed, extended with precise guidelines e.g. concerning the free testing length, and it should only be used to show that screws have enough deformation capability before breaking. If this is not wished for, then also precise information concerning the definition and measurement of the bending angle should be included. Finally, a decision must be taken at which bending angle the yield moment shall be determined and how to deal with the higher yield moment in the smooth shank area of partially threaded screws.

5.4 Insertion moment

The insertion moment M_{insert} is an important property that helps understanding the resistance of screws to being drilled into timber. M_{insert} is not needed for design; it only ensures that screws can be inserted without breaking. For certification purposes, the characteristic torsional moment capacity $M_{tor,k}$ must be at least 50% higher than the mean maximum insertion moment $M_{insert,mean}$ (EAD 130118-00-0603, 2016)[105]. During a test to determine M_{insert}, the relationship between measured insertion moment and insertion depth is recorded, and M_{insert} is the maximum value reached until the screw head touches the timber. The speed of the insertion process was always 100 rpm[106], and the screw with the longest threaded part was used. One series of usually 10 or 20 tests was carried out using the same piece of timber with a density higher than the mean density of the used species/strength class[107]. A proper selection with regard to density is important, as the insertion moment increases with increasing densities. If a screw can be inserted in spruce, this does not mean that it can also be inserted in beech LVL as well. To guarantee that no steel failures occur during insertion, a safe value for the insertion moment should be determined for each wood species/wood product and hence, used specimens must have high densities.

105 EN 14592 (2012) instead states that $M_{tor,k}$ / $M_{insert,k} \geq 1.5$; prEN 14592 (2017) corresponds to the EAD.

106 It is unknown if changes would influence results and what other testing bodies apply.

107 The EAD (2016) states that M_{insert} "has to be adjusted with the factor $k_{\rho E} = 480/\rho$, where ρ is the density of the test specimen, when softwood strength classes C16 - C40 or GL24 - GL36 are used. For other species, M_{insert} "has to be adjusted to the mean density of the respective material". EN 14592 (2012) refers to its "annex B", which is not there; prEN 14592 (2017) refers to EN 15737 (2009), which only states that the density must correspond to that of the strength class used in practice. Only annex B of the 2009 version of EN 14592 clearly states that all specimens must have a density between 400 kg/m^3 and 500 kg/m^3, implicitly assuming that only softwood is used.

The database contains 6863 individual test results in 524 test series, where 860 were tests with stainless steel screws and ten with hot-dip galvanised screws. Particularly long austenitic screws are critical, as these are not hardened[108]. All data where screw failures occurred were deleted. Except for ten tests at 30° between screw axis and fibre direction[109], all tests were done inserting the screws perpendicular to the grain. The used softwood products were 75.3% spruce (*Picea abies*), 4.1% spruce LVL and to a minor extent radiata pine (10 tests) and LVL made from radiata pine (20 tests). Except for 40 tests with predrilled spruce, softwood was not predrilled. Concerning hardwoods, most insertion tests used beech LVL (1005 tests) followed by beech solid timber (280 tests). Further hardwood species include oak (69 tests), ash (20 tests) and spotted gum (10 tests). Ten tests used non-predrilled oak and beech and 84 tests were executed using non-predrilled beech LVL. All other tests with hardwoods were predrilled. In total, 80.8% of all tests were done using timber with non-predrilled holes. The predrill diameters are shown in Figure 5-15 on the left. Concerning beech LVL, generally glued laminated timber (glulam) made from beech LVL was used. The screws were inserted into the secondary glue lines of the edge grain, i.e. the glue lines that result from the production of glulam. These secondary glue lines were observed to be the most punishing, which can be explained with surface densification of LVL (Frese, 2019). Only 20 tests were carried out where the screws were inserted in predrilled holes in the face grain. Of the 20 tests using radiata pine LVL, 10 were carried inserting the 8 mm screws in the face grain and 10 in the edge grain. Concerning tests in spruce LVL, the insertion direction is unknown in 80 of the 280 tests. In the other 200 tests, the screws were inserted in the face grain of spruce LVL.

Only few comparative test series were carried out, i.e. where timber was non-predrilled and predrilled (two screw types), where two different predrill diameters were applied (one screw type) and where the screws had shank ribs or not (one screw type). Deviations from the common test setup were recorded for 90 tests:

- 40 tests were carried out with screws with d_{nom} = 14 mm. The shortest 14 mm screws with a nominal length of 400 mm were directly inserted in non-predrilled spruce, whereas the 30 longer screws with nominal lengths of 1400 mm, 1500 mm and 2000 mm were inserted in short, 70 mm deep holes of 8 mm diameter ("pilot holes") in order to avoid "meandering" of these long screws.
- In 50 tests, 240 mm long screws were inserted only 64 mm deep in non-predrilled spruce.

---

108 In terms of certification, this is important to know, as no difference is made between carbon and stainless steel screws for the determination of the insertion moment if the screw geometry is the same. However, if long carbon steel screws are tested, this does not automatically mean that also the stainless steel screws with the same length can be inserted.

109 The only series at 30° comprised 10 mm screws inserted in the edge grain of beech LVL.

The insertion moment M_{insert} is influenced by different parameters. Many development efforts of screw producers, such as the addition of rib shanks for partially threaded screws, modified tip types and coating, aim at decreasing the resistance to insertion, which is particularly important for long screws. Looking at the whole dataset shown in in Figure 5-15 on the right, the density is certainly not the only factor influencing M_{insert}. The values larger than 65 Nm are insertion moments for 14 mm screws with a length of 1500 mm or 2000 mm. And, looking at the few data available for spotted gum with densities above 1000 kg/m^3, screws (d_{nom} = 8 mm) are available that have rather low insertion moments. Obviously, M_{insert} depends on the nominal diameter and length. Therefore, a "normalised" value can be calculated by dividing M_{insert} through d_{nom} and L_{nom}[110]. However, also using a "normalised M_{insert}" does not lead to a dependency on density; the general shape of Figure 5-15 on the right remains the same. This statement, however, only holds when looking at the *whole screw population*. If the same, individual screws are inserted in different wood species, obviously species with higher densities have a higher resistance against insertion. This can be seen in Table 5-4, where results of the few comparative tests are given. This means that trends, which are discernible in comparative test series where most parameters, i.e. the used screws, are kept constant, are not discernible anymore when looking at a much more comprehensive database, i.e. where many different screws are used.

Table 5-4 Results of comparative insertion tests.

	Spruce	Oak	Ash	Beech	Beech LVL
Screw 1: Partially threaded screw, d_{nom} = 8 mm, L = 400 mm, L_g = 100 mm, $d_{predrill}$ = 6 mm					
ρ_{mean} in kg/m^3	-	656	641	704	807
$M_{insert,mean}$ in Nm	-	4.7	5.2	5.5	6.9
Screw 2: Screw with two threaded parts, d_{nom} = 5 mm, L = 80 mm, $L_{g,tip}$ = 50 mm, $L_{g,head}$ = 15 mm					
ρ_{mean} in kg/m^3	450	594	-	728	-
$M_{insert,mean}$ in Nm	2.6	3.6	-	6.0	-

110 If plotting M_{insert} versus d_{nom} and L_{nom}, a dependency of M_{insert} on these two geometric parameters can be observed, which is reduced considerably when a normalised value for M_{insert} is used.

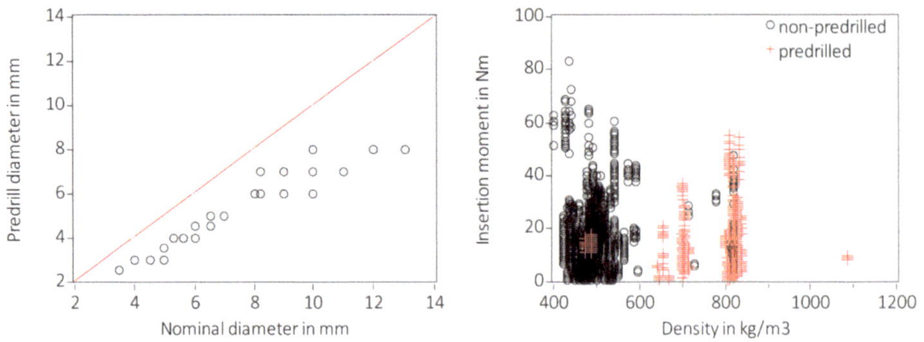

Figure 5-15 Left: Predrill diameter versus nominal diameter; bisect line is shown. Right: Insertion moment M_{insert} versus density (one piece of timber per series).

Concerning additional influencing factors, a detailed study concerning insertion moments and withdrawal capacities of screws inserted in predrilled and non-predrilled beech LVL was carried out by Frese and Blaß (2017). Frese and Blaß studied the insertion behaviour of different screw types by thoroughly examining insertion moment – insertion depth curves. They could show that, at least for species with high densities, screws with a normal tip, a small pitch, a large ratio between inner and outer diameter and an inorganic coating are best suited to obtain small insertion moments. This combination of thread geometry (small pitch and large ratio), however, led to small withdrawal capacities. A developed 8 mm prototype screw had a normal tip, a pitch of 5.6 mm and a ratio between inner and outer diameter of 0.68, which led to a good compromise between (small) insertion moment and (high) withdrawal capacity. For the database, the respective parameters are shown in Figure 5-16, where, contrarily to Frese and Blaß, insertion moment – insertion depth curves could not be examined. The ratio between inner and outer diameter (left) does not seem to influence M_{insert}, whereas a trend can be observed when looking at the pitch (right), with smaller M_{insert}-values for smaller pitch values. Furthermore, a slight trend of higher M_{insert}-values for screws with drill tips can be observed in Figure 5-16 on the right[111]. However, reconsidering the fact that M_{insert} depends on d_{nom} and L_{nom}, Figure 5-16 on the right is deceptive, and all influences vanish if again the "normalised" insertion moment is used instead of the measured value M_{insert}, see Figure 5-17 on the left.

Generally, the influence of geometric parameters such as pitch and ratio between inner and outer diameter cannot be properly investigated as long as screws of a certain nominal diameter are produced using always the same wire diameter. Screws are produced from wires, where the amount of steel material of the originally round wire remains the same after cold forming albeit in a different geometric form (thread). This means that e.g. pitch and ratio

[111] More investigations concerning the influence of the screw tip cannot be made as tips are not clearly defined geometrically.

between inner and outer diameter are correlated. For instance, a screw with a large ratio and a small pitch cannot be produced from the same wire as a screw with the same ratio and a large pitch; the amount of steel of the round wire and the final thread is too different. This leads to hidden correlations that cannot be discovered. For instance, analysis of insertion moments may reveal that a small pitch leads to small insertion moments. As for a certain nominal diameter, the pitch is correlated with the ratio between inner and outer diameter, the real influencing factor may be the ratio and not the pitch. Or rather, it is impossible to distinguish between both. Also thread flank angles and minor changes in tip geometry may have an influence, but were not recorded in the database.

The third parameter with a strong influence on M_{insert} identified by Frese and Blaß (2017), the inorganic coating, cannot be investigated here. If any information at all on coating is given in the database, then anti-friction coating is intended, which, according to Frese and Blaß, has less influence on M_{insert}. Therefore, even if a coating is mentioned in the reports, it is usually not specified which type of coating it is and for 87% of all insertion tests, no information on coating is given. A known important influencing factor on M_{insert} was hence not recorded and obviously cannot be analysed; rather, its effects even risk to be assigned to other parameters.

Consequently, the database does not help to propose optimised screw geometries that help keeping insertion moments low whist reaching high withdrawal capacities.

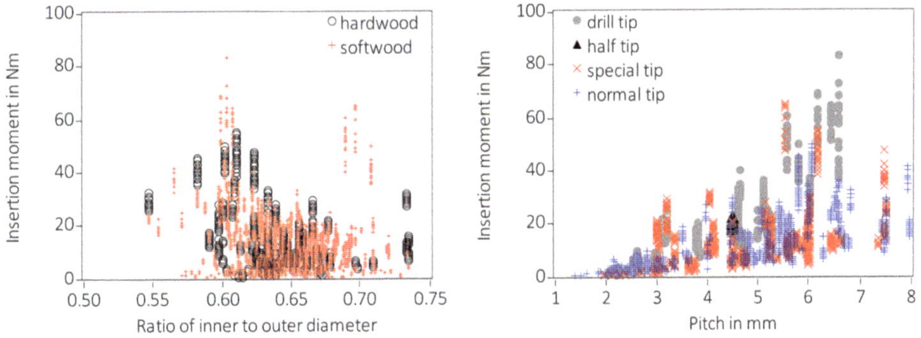

Figure 5-16 Left: Insertion moment M_{insert} versus ratio of inner divided by outer diameter (6763 individual tests). Right: Insertion moment M_{insert} versus pitch (5593 individual tests).

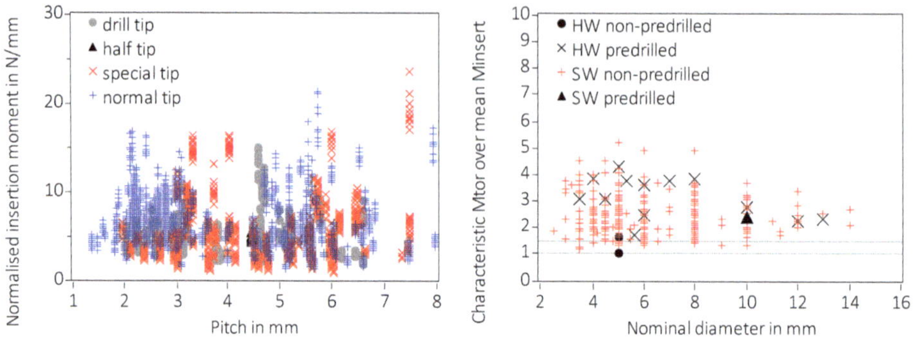

Figure 5-17 Left: "Normalised" insertion moment versus pitch (5593 individual tests). Right: Ratio of characteristic torsional moment capacity over mean insertion moment; HW = hardwood, SW = softwood.

For certification purposes, M_{insert} is compared with M_{tor} in order to guarantee that screws can be drilled in without breaking. Within the database, 292 series are available where both M_{insert} and M_{tor} were measured. Figure 5-17 on the right shows the ratio of the characteristic torsional moment capacity $M_{tor,k}$ divided by the mean insertion moment $M_{insert,mean}$. $M_{tor,k}$ was calculated in accordance with EN 14358 (2016), based on a lognormal distribution and taking the number of tests per series into account. In terms of certification process in accordance with the EAD, the horizontal line at 1.5 marks the boundary between "pass" and "fail". 13 series (4%) failed, where the lowest ratio of 1.0 was a series using non-predrilled beech (ρ_{mean} = 728 kg/m^3, with 80 mm long 5 mm screws with two threaded parts)[112]. With regard to practical implications, i.e. that screws must not break during the insertion process, it may be recommendable to use 95th percentiles (Hübner, 2013) or even maximum values of M_{insert} (Frese, 2019) to compare with $M_{tor,k}$. This is of particular importance seeing the lack of clear regulations concerning the selection of the timber pieces with regard to density, see also footnote 107.

Finally, a further issue that has to be dealt with in practice has not been addressed up to now. Currently, screws are inserted in timber, where the longest screws must be used, and determined insertion moments are valid for timber classes with a certain density. Screws are however also used in steel-to-timber joints, where different stresses act on the screws at the end of the insertion process, i.e. when the screw heads touch the steel plate.

[112] The second black dot in Figure 5-17 on the right is the same screw, but in non-predrilled oak with ρ_{mean} = 594 kg/m^3

5.5 Withdrawal capacity

5.5.1 General

The withdrawal capacity F_w is especially important for joints with screws, because these joints are, above all other joints with dowel-type fasteners, perfectly suited to transmit high tensile forces in axial direction of the fastener due to the high withdrawal resistance of screws[113]. Especially joints with inclined screws are efficient, as the high axial stiffness and capacity of screws are activated. In addition, joints with laterally loaded screws show increased capacities due to a high rope effect contribution of the screws. Consequently, a considerable amount of publications is available dealing with withdrawal behaviour of screws, with publications by Brandner (2019), Ringhofer (2017), Brandner et al. (2018), (2019) and Westermayr and van de Kuilen (2020) being the most recent. A particular study was carried out in Hölz (2021), who investigated the influence of pitch and thread flank angle on the withdrawal behaviour with the aim to develop an optimised thread geometry.

Although Eurocode 5 (EN 1995 1-1, 2010) gives an equation to calculate a characteristic withdrawal parameter $f_{ax,k}$ (used to calculate the withdrawal capacity), usually, technical documentation of screw producers is consulted to obtain $f_{ax,k}$-values. In the framework of certification testing, withdrawal tests are carried out in accordance with EN 1382 (2016)[114]. The screws are inserted at an angle α to the grain and with a penetration length of L_p "of the threaded part"[115]. The timber pieces are large enough to avoid splitting and the screw tips are not protruding. No deformations are measured, only the maximum force during withdrawal and the density of the timber pieces is determined.

In total, 7632 individual tests grouped in 579 series were assembled in the database. All test results with failures other than withdrawal failures were deleted. 74.8% of the tests were carried out with an angle $\alpha = 90°$ between screw axis and grain direction, 12.8% with $\alpha = 0°$ and 6.1% with $\alpha = 45°$. The rest was carried out with angles of 15° (260 tests), 30° (80 tests), 60° (40 tests) and 75° (100 tests), where not all species were used at all angles. In general, half of the tests per series were inserted in radial and the other half in tangential direction if $\alpha = 90°$. Spruce (*Picea abies*) was used for the majority of tests (52.6%), either solid timber or glued laminated timber (no differentiation in the database), and 28.7% of the tests were carried out using beech LVL. Other species and products include oak (4.3%), beech (2.3%)

[113] Withdrawal capacities alone are not sufficient for design; the capacity of joints with axially loaded screws (in tension) is a function of withdrawal, tensile and, if applicable, head pull-through capacity.

[114] Respectively older versions of the same standard, depending on the date of test execution. Both EN 14592 (2012) and the EAD (2019) refer to EN 1382, where the EN requires 10 withdrawal tests and the EAD 20.

[115] In the assembled database, the penetration length including the screw tip is given. Contrarily to nails, a screw tip length is not measured in the framework of certification testing.

and LVL made of radiata pine (1.6% or 120 tests). Fewer results are available for radiata pine, spotted gum, LVL made of spruce and ash (less than 0.8% each). In addition, wood-based panels (OSB 5.2%, particleboard 2% and SWP 0.5%) were used, where the screws were drilled through, i.e. where the screw tip was outside the panel. In all other tests, the screw tips were embedded in the timber. The penetration length L_p differed as can be seen in Figure 5-18 on the left, where L_p ranged between $4 \cdot d_{nom}$ and $12 \cdot d_{nom}$ for the standard withdrawal tests with the screw tip embedded in timber. The tests with wood-based panels were all drilled through (red crosses in Figure 5-18 on the left) and L_p was equal to the panel thickness. Figure 5-18 on the right shows that the penetration length L_p was not always shorter than the length L_g of the threaded part. The series oscillating around 1.0 were tests with short screws, where L_p was more or less L_g (the series with a ratio of 1.05 were 6 mm screws with L_g = 66.5 mm and L_p = 70 mm). The three series with ratios of 1.58, 1.36 and 1.27 used screws with significantly shorter thread length L_g than penetration length L_p. Nominal diameters ranged from 3 mm to 14 mm, with 6 mm (21%), 8 mm (21%), 10 mm (14%) and 5 mm (12%) being the most frequent.

In general, partial insertion of the smooth shank of partially threaded screws into the timber by a length l_{emb} (Pirnbacher et al., 2009) helps increasing the withdrawal capacity, as the threaded part, that transmits forces, starts only at a depth l_{emb} from the timber surface, leading to a larger stressed timber volume. This procedure was not applied for any of the tests contained in the database.

Concerning tip types, withdrawal tests with all four tip types specified in Table 5-1 were carried out. Most screws had a normal tip (46%) and 42% had a special tip. Drill tips were present in 12% of the screws (of which 75% were fully threaded screws) and only one series with 10 tests used a screw with a half tip. This differentiation is important as drill tips usually do not have a withdrawal resistance; see also Figure 5-21, where the drill tips protrude from failed test specimens. The ratio of inner over outer diameter was ranging from 56% to 73%, see also Figure 5-2 on the left.

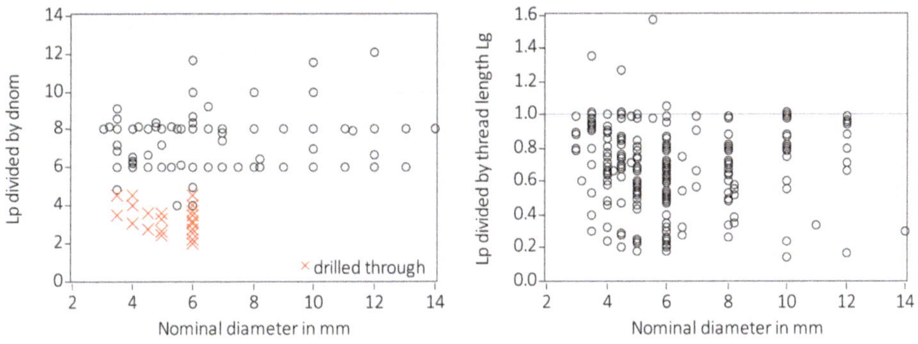

Figure 5-18 Left: Ratio of penetration length L_p divided by nominal diameter d_{nom}; 579. 'Drilled through' are tests with protruding screw tips. Right: Ratio of penetration length L_p divided by length of the threaded part L_g; 6015 values.

Concerning steel grades, 474 of the 7632 screws were made of stainless steel; all other were of carbon steel. The reason for this imbalance is that withdrawal parameters of carbon steel screws are valid also for stainless steel screws if tip type and thread geometry is the same. As many screw producers have harmonised geometries for carbon and stainless steel screws, only screws of one steel grade, i.e. carbon steel, were tested.

Only few comparative series were carried out, with which specific influencing factors were investigated. Figure 5-19 shows the results of a test programme, where a possible time dependency of the withdrawal capacity was investigated, by leaving certain time spans between insertion and withdrawal. No difference can be seen in the results. A further series looked into the difference between a screw with different tip types[116]. Finally, the influence of predrilling and predrill diameter in beech LVL was investigated, see also Table 5-6 and Figure 5-32 in section 5.5.4.

116 12 tests each with d_{nom}= 6 mm, L_p = 48 mm, ρ = 456 kg/m^3: $F_{w,normal\ tip}$ = 3.6 kN and $F_{w,special\ tip}$ = 5.3 kN

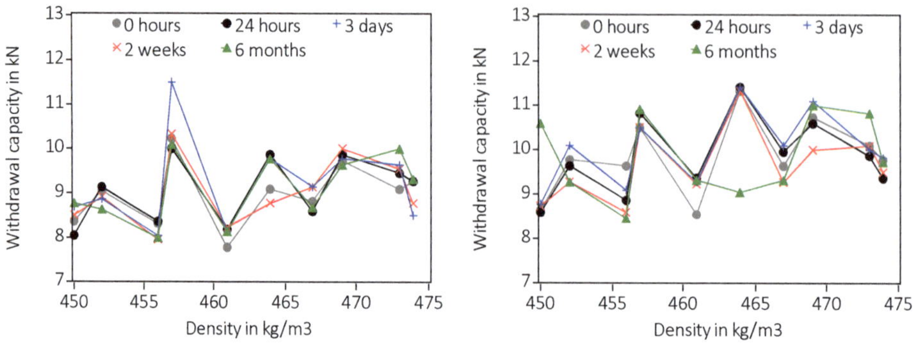

Figure 5-19 Withdrawal capacity F_w versus density ρ during test, screws withdrawn at indicated time spans after insertion, using screws with d_{nom} = 8 mm and L_p = 64 mm. Left: Screw 1 with normal tip. Right: Screw 2 with special tip.

An important parameter is not only the withdrawal capacity, but also the withdrawal stiffness. Recent publications (Brandner et al., 2018; Ringhofer, 2017) address specifically stiffness, because in particular joints with inclined screws can reach significantly high stiffness and capacity values and can be used e.g. to design mechanically jointed beams. For such structures, stiffness values are required. However, uncertainties are high concerning withdrawal stiffness, which is above all due to the dependency of stiffness values on the test setup (Blaß and Steige, 2018) and the difficulty of measuring very small deformations, or, as Dietsch and Ringhofer (2021) stated, the lack of clear definitions about how to determine withdrawal stiffness. Concerning the assembled data, no information on the stiffness is given, because deformations were not measured during testing.

5.5.2 Determination of the withdrawal parameter

Measured withdrawal capacities are a function of various parameters, e.g. nominal diameter d_{nom} and penetration length L_p[117], where a normalisation must be carried out, in order to allow for comparison of test results. A common simple model assumes a uniform shear stress distribution along the screw axis and, analogously to nails, the withdrawal capacity F_w is normalised by dividing F_w through the nominal diameter d_{nom} and the effective penetration length L_{ef} in accordance with Eq. (4-10). This leads to a conventional value of the withdrawal parameter f_w in MPa, which can then be analysed further to investigate other influencing factors such as the density ρ. The term ***conventional*** is of importance here, as obviously, the thus calculated withdrawal parameter f_w does not deliver a correct "strength" value:

[117] L_p in turn is a function of d_{nom}, see Figure 5-18 on the left.

- The area to be considered when calculating a withdrawal "strength" should be the surface area $\pi \cdot d_{nom} \cdot L_{ef}$, because the screws are withdrawn along the outer diameter $d_o = d_{nom}$, and not the projected area $d_{nom} \cdot L_{ef}$.
- The underlying model may not be true, as the stress distribution along the axis can be assumed to be non-uniform depending on the penetration length L_p, see also Figure 5-20 on the left.
- It is difficult to determine an effective penetration length L_{ef} (i.e. subtracting the screw tip) for screws with other than drill or half tips. Even for screws with drill tips, the determination of a tip length is not trivial, see also Figure 5-22 on the right.

Projected versus surface area

Currently, certification test results and values given in technical documentation and thus the order of magnitude of f_w-values are based on the projected area $d_{nom} \cdot L_p$, where the difference with the surface area is π. Hence, the procedure is not changed here only for the sake of a modification by a constant factor of 3.14. There are approaches however that require f_w to be calculated using the surface area. Westermayr and van de Kuilen (2020) for instance relate the withdrawal capacity of screws inserted parallel to the grain with the longitudinal shear strength. Here, obviously, the correct stress surface is needed and hence, the surface area must be considered. Also for tests with screws inserted at angles to the grain, such relations are thinkable, correlating e.g. withdrawal of screws inserted perpendicular to the grain with the transverse shear strength (see also section 4.9.7 of Hübner, 2013).

Stress distribution

It is difficult to assess the influence of the penetration length L_p on the withdrawal stress distribution along the screw axis without bespoke test series[118]. In addition, finite element modelling would enhance knowledge, but no precise enough models exist that are able to model screws with a thread inserted in (non-predrilled) timber and loaded in withdrawal. Often, simplified models using cohesive surfaces and damaged zones with lower properties to model the withdrawal behaviour are applied (Azinović and Frese, 2020). Other approaches model screws as cylinders inserted into a tube representing the system of wood and screw thread, where the tube's shear stiffness is representative of the axial stiffness (Dietsch et al., 2019; Mestek, 2011). In addition, the penetration length as a function of the nominal diameter is frequently kept constant during testing. Various researchers looked in detail on how L_p influences the withdrawal capacity F_w. Especially Stamatopoulos and Malo

[118] There is a limit in such test series however; large penetration lengths cannot be chosen as then screws will fail in tension.

(2015), Pirnbacher et al. (2009) and Ringhofer (2017) performed bespoke test series, which, in the case of Stamatopoulos and Malo, were dealing with threaded rods and included a theoretical approach based on the Volkersen theory[119]. Ringhofer and Schickhofer (2014) could confirm the suitability of the Volkersen theory to predict the withdrawal capacity. They carried out withdrawal tests using several strain gauges along the screw axis to measure the stress distribution, and compare experimental and analytical results. All authors concluded that the withdrawal properties depend linearly or slightly non-linearly on the penetration length, i.e. a quasi-linear stress distribution. Recently, Claus et al. (2022) presented a new testing method using fibre Bragg gratings to measure the stress distribution along the axis of fully threaded screws during withdrawal. They found a dependency of this method on load level, load-to-grain angle (= screw axis-to-grain angle), density and support conditions. The results indicate that stress distributions are nonlinear and depend on the penetration length. The strain distribution along the screw axis instead was measured by Kumpenza et al. (2020) using optical measurements combined with numerical simulations. They concluded that strains are maximum at the screw insertion point on the timber surface and whereas shear strains are nearly linearly distributed over the insertion length, the transverse strain decrease rapidly versus the screw tip. Kumpenza et al. did not investigate different insertion lengths, their findings, however, indicate, that not all regions of the screw thread contributed equally to the withdrawal.

Existing models to calculate the withdrawal capacity/parameter of screws were usually derived via regression analyses based on test results. Only few models are available, where L_p was chosen as independent variable for the regression. For instance, the model used in the current version of Eurocode 5 (Eq. (8.39) in EN 1995 1-1, 2010) gives a dependency of the withdrawal parameter on the penetration length to the power of -0.1. This means that the withdrawal parameter is weakly decreasing with increasing L_p. Another model, based on data with variable L_p, chose a polynomial model approach, so that no clear statement can be given concerning the influence of L_p (Frese et al., 2010).

To add to this discussion, the influence of L_p on the withdrawal parameter is investigated qualitatively with the assembled withdrawal data. Figure 5-20 on the left shows the withdrawal parameter f_w versus the penetration length L_p. for all tests at 90° with spruce. No correction was carried out to eliminate the influence of density differences on f_w. It can be seen that the minimum values are almost constant over the whole range of L_p, whereas the maximum values tend to decrease with increasing L_p. As less testing data is available for longer L_p, it is difficult to say if this observed trend is a real observation or a statistical artefact. Additionally, contrary to Pirnbacher et al. (2009), f_w is not increasing with longer L_p, where here, however, the variety in screw types and timber density is much larger and may lead to the scattering effects observed in Figure 5-20 on the left. To conclude, the shape of the stress distribution along the screw axis is not easy to assess, also because comprehensive

[119] Stamatopoulos and Malo assumed a rather steep load distribution inside the timber of 1:3.

studies comprising significantly different penetration lengths are difficult to carry out (see footnote 118), i.e. in particular investigations with large penetration lengths are missing. However, it can be safely assumed that the stress distribution will be the more nonlinear the longer the penetration length, or, as Ringhofer (2017) puts it, significant nonlinear stress distributions seem to occur only at penetrations lengths that do not lead to withdrawal anymore, but to tensile failures of the screws.

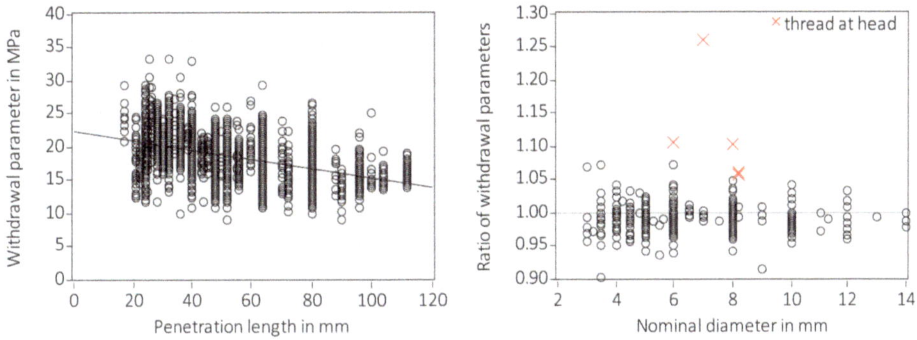

Figure 5-20 Left: Withdrawal parameter f_w (using d_{meas}) versus penetration length L_p for spruce at 90°; 3447 values, trendline is shown. Right: Ratio of withdrawal parameters: Parameter calculated with nominal diameter divided by parameter calculated with measured diameter; 7632 values. 'Thread at head' are tests with screws with two threaded parts, where the thread close to the screw head was withdrawn.

Effective penetration length L_{ef}

Finally, an ongoing discussion is the definition of the effective penetration length L_{ef}, where in the case of nails, the nail tip is subtracted from the full penetration length (see Eq. (4-10)). For screws however, the length of the tip is difficult to assess, see also Table 5-1 and Figure 5-3, where for instance normal and certain special tips do not have a well-defined length. Furthermore, tip lengths are not measured nor is this information available to practitioners (drill tips excluded). In addition, bespoke test series must be carried out to assess the influence of the tip length; e.g. as performed in Hübner (2013), who performed tests with drilled-through and embedded tips. He found a difference in withdrawal capacity between both series, with higher capacities for the drilled-through series, and could therefore calculate a correction factor that adjusted the capacities of the tests with embedded tips to those with drilled-through tips. However, only screws with half tips and special tips similar to drill tips were used, where a more prominent influence of the screw tip can be expected in comparison to a normal tip. A proposed approach to define L_{ef} by subtracting the tip length is to subtract a percentage of the nominal diameter; $L_{ef} = L_p - \{1.11 \div 1.17\} \cdot d_{nom}$ (Hübner, 2013; Pirnbacher et al., 2009), which reduces the adjustment to a modification by a constant factor. To further investigate this issue, a small test series was carried out that is presented and discussed in section 5.5.3.

Generally and in accordance with EN 1382 (2016)[120], screw tips are embedded in the timber during withdrawal tests and tip lengths do not form part of the geometrical properties to be measured. Consequently, no analyses can be carried out with the database concerning the influence of the screw tip. Withdrawal parameters stemming from certification tests are hence calculated including the tip length; $L_{ef} = L_p$, where, however, any smooth shank parts must be subtracted, see Figure 5-18 on the right.

Conclusion

The withdrawal parameter is a ***conventional*** value, which allows for the calculation of the withdrawal capacity of a screw with a certain nominal diameter and a certain penetration length. Modifications to obtain mechanically correct (strength) values by considering the surface area change this conventional value by a constant factor. There is no scope to change the definition of f_w in this study, which can be easily done in further research by adding a "π" in the calculation sheets. Here, the currently used definition is used, i.e. a withdrawal parameter and not a withdrawal strength, because it delivers values whose orders of magnitude are familiar.

For the analyses however, more "realistic" withdrawal parameters can be considered if instead of the nominal diameter d_{nom} the measured outer diameter d_{meas} is used in Eq. (4-10). In such a case, regressions with measured data per individual test are possible, and no nominal data is required. Figure 5-20 on the right shows that the difference in withdrawal parameters calculated with nominal and measured diameters scatter by approximately ±10%. If screws have two threaded parts, tests with both threaded parts, the thread close to the tip and close to the head, are carried out. The red plusses show the data for tests with the thread close to the screw head, whose outer diameter is considerably larger than the nominal diameter of the screw, which explains the large ratios.

Therefore, for all analyses carried out in this chapter, the withdrawal parameter f_w is calculated by dividing the withdrawal capacity F_w through the full penetration length L_p of the threaded part[121] (including screw tip) and the measured nominal diameter (= outer diameter) d_{meas}[122].

120 Also in the 1999 version of EN 1382, the screw tip is embedded in the timber and not drilled-through.

121 f_w of the three outliers in Figure 5-18 on the right with $L_p/L_g > 1.2$ were calculated using the (shorter) thread length L_g instead of the (longer) penetration length L_p.

122 In 371 cases, the outer diameter was not measured; here $d_{meas} = d_{nom}$.

5.5.3 Additional test series

Independently of what was said in the previous section, i.e. that the full penetration length is considered when calculating f_w, screw tips will influence the withdrawal capacity. In particular, drill tips (and certain special tips) will provide no resistance to withdrawal, as can be seen exemplarily in Figure 5-21 that shows the failure area of a softwood beam with a reinforced hole. The drill tip of the reinforcing screw did not provide any resistance because the failure plane (tensile failure perpendicular to the grain at the hole) was at the level of the threaded area, and not at the screw tip. Therefore, a small additional test series was carried analogously to the tests with nails presented in section 4.5.2.

Here, two 8 mm screws were withdrawn from five pieces of timber, inserting eleven screws per timber piece with eleven different penetration lengths along the fibre direction. The chosen penetration lengths were {2, 3, 4, 5, 6, 7, 8, 9, 10, 11, 12} · nominal diameter d_{nom}. This means that per penetration length, five test results are available. The chosen screws differed in their thread geometry and above all in the tip type; one screw had a normal tip and the other a drill tip. Photos of the used screws are shown in Figure 5-22. The geometric data of the screws were measured, and the measured tip length of both screws was approximately equal to the nominal diameter of 8 mm, see Figure 5-22. The densities of the timber pieces were 374, 390, 427, 456 and 466 kg/m^3 at moisture contents between 12.5% and 13.5%. Obviously, only two screws were used, impacting on the representativeness of the data.

Concerning minimum penetration lengths in accordance with Eurocode 5, the situation is not very clear for laterally loaded screws. As the rules for nails apply for screws with $d_{nom} \leq 6$ mm, the minimum penetration length for these screws results to $6 \cdot d_{nom}$. For screws with larger diameters, the rules for bolts apply and hence, for obvious reasons, no minimum penetration length is given. For axially loaded screws instead, Eurocode 5 prescribes a "minimum pointside penetration length of the threaded part" of $6 \cdot d_{nom}$. European Technical Assessments instead often contain detailed rules concerning minimum values of penetration length or timber thickness, where often, minimum penetration lengths of $6 \cdot d_{nom}$ can be found (for screws inserted perpendicular to the grain). The penetration length of the tests contained in the database ranged from $4 \cdot d_{nom}$ to $12 \cdot d_{nom}$, see Figure 5-18 on the left (all except tests with wood-based panels).

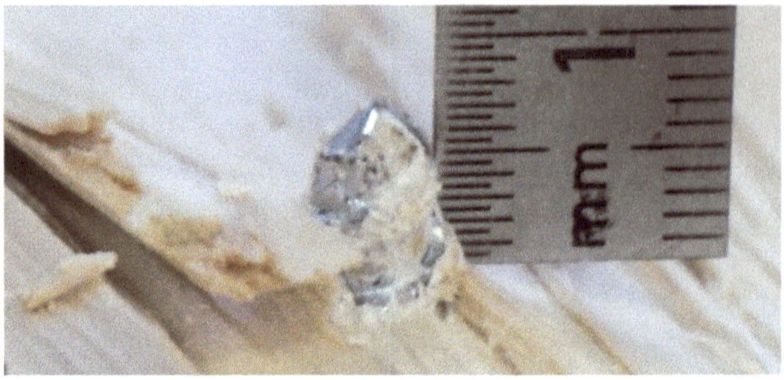

Figure 5-21 Drill tip protruding from test specimen failed due to tensile loading perpendicular to the grain.

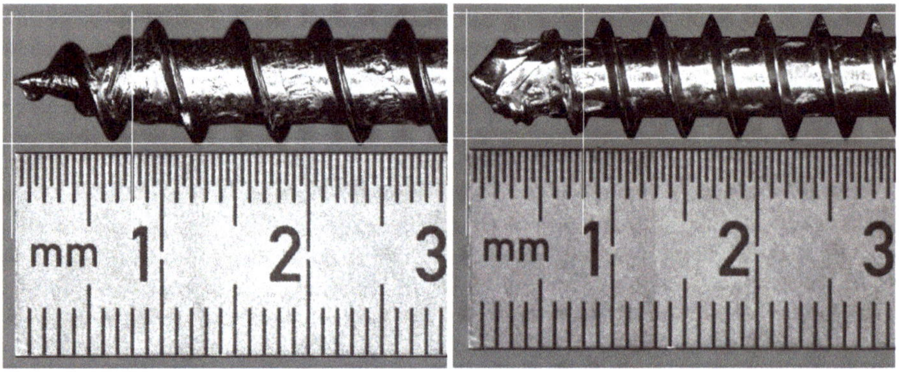

Figure 5-22 Detail of tip area of used screws with nominal diameter d_{nom} = 8 mm; a tip length of 8 mm is visualised (Nota bene: the thread at the tip goes around the screw). Left: Normal tip. Right: Drill tip.

The determined withdrawal capacities were normalised applying Eq. (4-10), i.e. subtracting the tip length from the penetration length ($f_{w,Lef}$). The capacities were also normalised applying Eq. (4-11), i.e. using the full penetration length ($f_{w,Lp}$). In Figure 5-23, the results are presented. In comparison to the results for nails shown in Figure 4-30 on the left, the results per density scatter less, with the exception of two high values for $f_{w,Lef}$ of the screw with the drill tip that was inserted at only $2 \cdot d_{nom}$ = 16 mm. However, also here, smoother curves, with less peaks, can be observed for the lower densities. Confirming the findings from Hübner (2013), a subtraction of the tip length leads to horizontal curves, with no influence of the penetration length on the withdrawal parameter, see Figure 5-23 on the left. This holds for screws with both tip types and is different to the findings for nails, see Figure 4-30 on the bottom. For penetration lengths $\geq 6 \cdot d_{nom}$, the curves show an asymptotic trend, where the screw with a drill tip shows, similar to the results for nails, a flatter curve than the screw with a normal tip, whose mean values continue to slightly increase at higher penetration lengths. In terms of numerical values, obviously, $f_{w,Lef}$ and $f_{w,Lp}$ will always differ by the percentage of the tip length on the penetration length (see also discussion around Figure 4-29 on the right). However, at penetration lengths used in practice, this percentage is low. It is more important to be consistent and to derive withdrawal parameters (and backwards calculate withdrawal capacities) always following the same procedure. The additional tests show that the currently used minimum penetration length of $6 \cdot d_{nom}$ is a good choice.

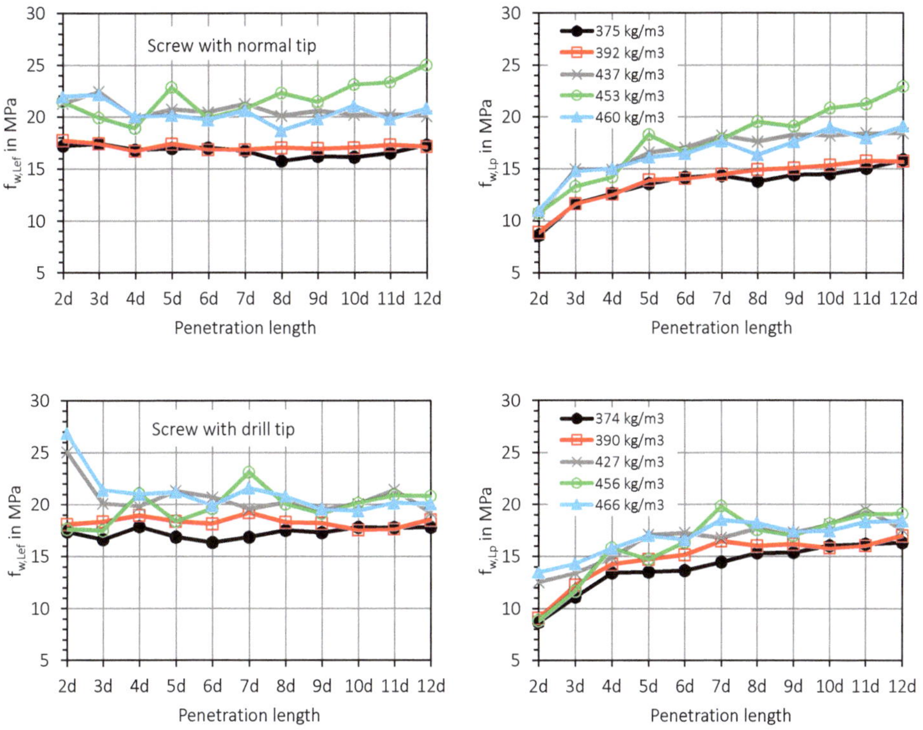

Figure 5-23 Withdrawal parameter versus the full penetration length. Top: Screw with normal tip. Bottom: Screw with drill tip. Left: $f_{w,Lef}$ calculated with the effective penetration length L_{ef} (subtracting the tip). Right: $f_{w,Lp}$ calculated with the full penetration length L_p (including the tip).

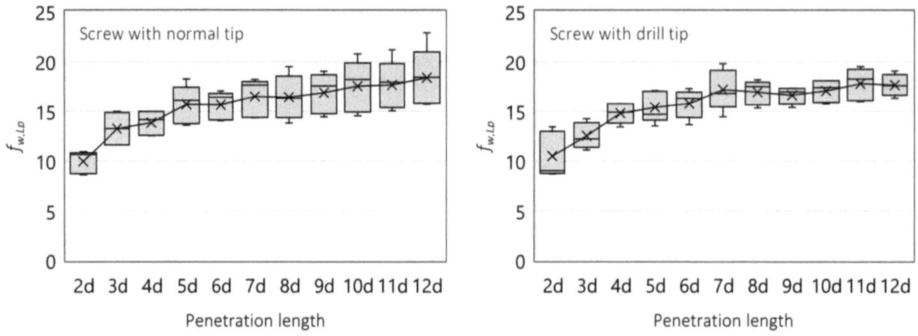

Figure 5-24 Withdrawal parameter $f_{w,Lp}$ versus the full penetration length. Left: Screw with normal tip. Right: Screw with drill tip.

5.5.4 Data subsets

An exhaustive regression analysis including simulation of empirically represented data was carried out on the yield moment of nails (see section 4.4). A major drawback in section 4.4 was the fact that the regression analysis tried to link observations that were not determined using the same nails (i.e. two different tests for M_y and F_t). This is different here, because factors influencing withdrawal capacities were evaluated in a single test. Therefore, a fundamental prerequisite of any regression analysis, the direct relationship between dependent and independent parameters, is not violated. Further issues discussed in section 4.4.1 arise also here, especially (i) if the database is representative enough, (ii) if relationships cannot be found because data is missing or (iii) if spurious relationships are found.

In addition, new challenges develop. Contrary to section 4.4, where the shape and variables of the regression equation were clear as the equation was based on the mechanical equation to calculate the full plastic bending capacity of a circular section, a possible regression approach must be carefully chosen here. Contrary to many studies available in the literature, the assembled database is not based on tests following a tailor-made test programme nor are all possible influencing factors recorded. Furthermore, the database is diverse, comprising many subsets that are distinguished by e.g. species or angles α between fibre direction and screw axis, and where clusters exist, with certain subsets being overrepresented.

The first task, therefore, is to divide the database into subsets that are strictly necessary to develop a meaningful regression approach or, in other words, that enable a decision which shape the regression equation should take and which independent variables should be considered. A good starting point is to look at the dependency of withdrawal properties on the angle α between screw axis and fibre direction and the density ρ in order to decide on subsets. In the literature, agreement exists on this relationship between f_w ,α and ρ, and all proposed regression equations are based on the density as one of the most important independent variables and they are differentiated by the angle α (Brandner, 2019; Frese et al., 2010; Ringhofer, 2017). Usually, also the nominal diameter comes into play.

Angle α between screw axis and fibre direction

If considering the findings from literature, it becomes clear that a differentiation into two groups with different angles α may be meaningful. Brander (2019) proposes 30° as the inflexion point of a bilinear model. Frese et al. (2010) stated that the load-carrying mechanism for screws inserted at $\alpha \geq 45°$ is principally different from that for screws inserted parallel or near-parallel to the grain[123]. Therefore, Figure 5-25 on the left shows all f_w-values for $\alpha \geq 30°$ and on the right, all data for $\alpha \leq 30°$ is given[124]. The differentiation shown in Figure 5-25 does not give clear results over the whole bandwidth of densities concerning the influence of α. Westermayr and van de Kuilen (2020) reported that an increase in withdrawal strength with increasing screw axis-to-grain angle can be observed for spruce, but not for beech. Therefore, the data is investigated further, looking at different wood species. Figure 5-26 on the left shows the same data as Figure 5-25 on the left, but now differentiated by timber product. The cluster with low f_w-values between 600 and 700 kg/m^3 stem from tests with wood-based panels[125] and $\alpha \geq 30°$. If now the results for spruce differentiated by the angle between screw axis and fibre direction is considered, Figure 5-26 on the right, the trendlines for α indicate that f_w-values increase with increasing α, confirming the findings of Westermayr and van de Kuilen. Figure 5-27 on the left shows the test results for beech LVL, where angles α of 0° and 15° lead to lower f_w-values than angles α of 45°, 75° and 90° (see also Figure 5-39 on the right), and where for $\alpha = 0°$, no correlation could be found. Figure 5-27 on the right instead shows the results for oak using 16 different screws[126], where the oak was predrilled except for 11 tests using screws with two threaded parts. The trendlines underline that also for oak, f_w-values increase with increasing α. For other hardwood products than beech, therefore, the findings of Westermayr and van de Kuilen cannot be confirmed.

---

123 Data at angles α between 0° and 45° were not available in 2010.

124 Data for $\alpha = 30°$ is hence shown two times.

125 More specifically, with OSB and particleboard; the results for SWP lay in the range of those for spruce.

126 For beech and ash, only tests at $\alpha = 90°$ were carried out.

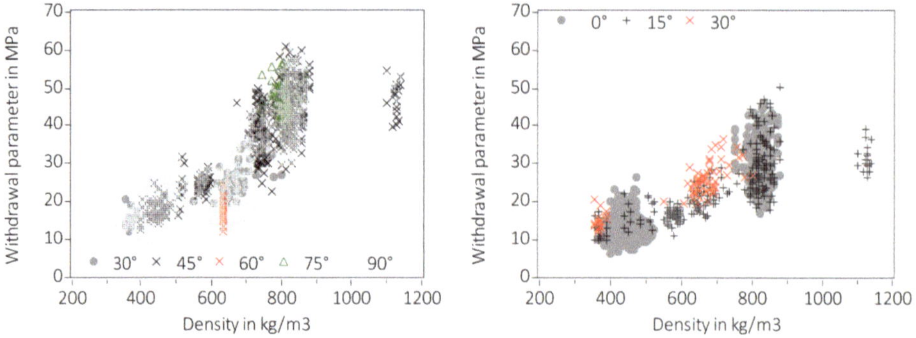

Figure 5-25 Withdrawal parameters versus density differentiated by angle α between fibre direction and screw axis. Left: Tests at $\alpha \geq 30°$, 6398 values. Right: Tests at $\alpha \leq 30°$, 1314 values.

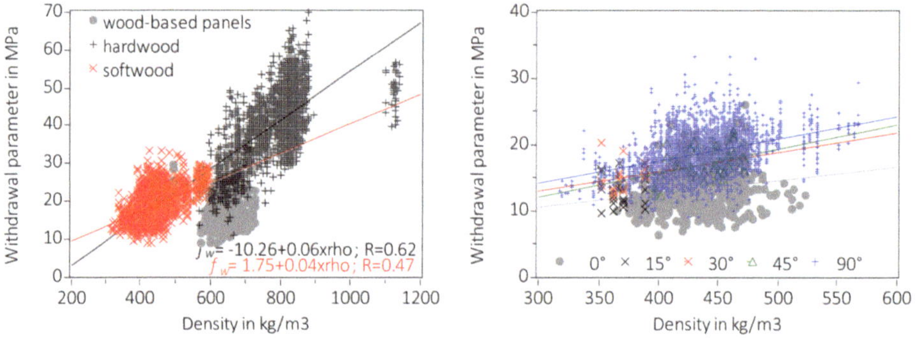

Figure 5-26 Left: Withdrawal parameters versus density differentiated by timber product. Tests at $\alpha \geq 30°$. Right: All tests for spruce differentiated by angle α. Coloured lines are trendlines of respective α.

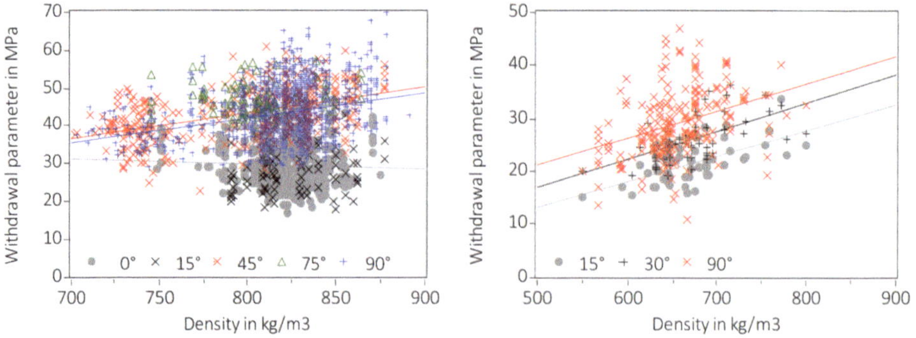

Figure 5-27 Withdrawal parameter versus density differentiated by angle α. Left: Data for beech LVL, 2193 values, trendlines are shown for $\alpha = 0°$, 45° and 90°. Right: Data for oak, 331 values.

Density ρ

Figure 5-26 shows that data for OSB and particleboard are significantly lower than data for solid wood, glued laminated timber and laminated veneer lumber made of softwood or hardwood species. The (extrapolated) trendlines, their equations and Pearson's correlation coefficients for softwood and hardwood (including spotted gum) are shown in Figure 5-26 on the left. For $\alpha \geq 30°$, the trend discussed by Brandner (2019) of a steeper increase of f_w-values with increasing density of hardwoods than of softwoods is confirmed. The small dataset with spotted gum with ρ_{mean} = 1123 kg/m^3 influences the inclination angle, i.e. excluding it increases the inclination angle for hardwoods even more. Due to its small size, it may be that scatter is not representative, see also Figure 5-38 on the left, where the influence of the small dataset with spotted gum is evident. Giving a closer look by further differentiating in lower density hardwood species, Figure 5-28 on the left shows that no difference can be observed between species ash, beech and oak for α = 90°. The results stem from tests using 21 different screws, where only five screws were used in comparative series, i.e. using the same screw in combination with different species. If now looking at numerical values of these bespoke test series, Table 5-5, withdrawal properties do increase with increasing density. This highlights difficulties around the determination of general design equations, which, considering the results of the comparative series given in Table 5-5, could incorporate higher values for beech, whereas Figure 5-28 on the left suggests that such differences may vanish in the population of all screws.

All data[127] for higher-density hardwood stem from beech LVL and not from solid wood/glued laminated timber and for these, the whole situation is more complex, as more parameters differ. Figure 5-28 on the right shows data for beech LVL differentiated by the insertion direction. No differences can be seen, and the highest f_w-values at densities of ca. 870 kg/m^3 are all determined using non-predrilled beech LVL. Therefore, it is not clear if the high f_w-values stem from insertion in secondary glue lines of the edge grain or if simply the tested screws work particularly well in non-predrilled beech LVL. Frese (2019) states that all relevant faces should be tested and that in the edge grain, tests should be performed also in the middle of the lamination (hence not in the secondary glue line), where lower f_w-values are expected in comparison to insertion in or close to the secondary glue lines. For all results with screws inserted in the edge grain, the exact location of the screw is however unknown.

The insertion direction however is not the only parameter that differs. Further differences encompass predrilling and predrill diameters (see Figure 5-31) as well as ratios of inner divided by outer diameter of the used screws (see Figure 5-36). Especially modern screws optimised for hardwoods have higher ratios, i.e. larger inner diameters in comparison to the outer diameters to increase their tensile capacity.

127 With the exception of spotted gum.

Seeing the persistent large scatter even for a highly homogenised product such as beech LVL, all data could be considered as one population to avoid many subsets that may not yield much[128]. Figure 5-29 on the left shows again the data for $\alpha \geq 30°$, but without the values for OSB, particleboard and spotted gum. The linear regression equation is valid for the whole population and is very similar to the equation for hardwood given in Figure 5-26 on the left, albeit with a much higher Pearson's correlation coefficient. At the first glance, it seems that no differentiation between softwood and hardwood is necessary, because any observed differences vanish in the scatter of the results. Such a possibility would facilitate design by reducing the complexity and amount of equations. However, it would mean that e.g. f_w-values for softwood with lower densities are underestimated, whereas the values for higher densities are overestimated. Figure 5-29 on the right shows the data for softwood and for $\alpha \geq 30°$, differentiated by species. A comparison with the trendline shown in Figure 5-29 on the left makes clear that withdrawal parameters for radiata pine LVL and high-density spruce would be overestimated if one global regression equation would be chosen. This differentiation is further discussed in section 5.5.6.

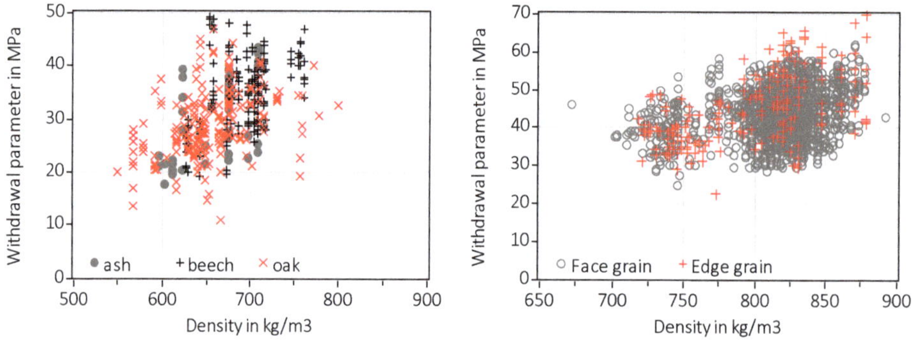

Figure 5-28 Withdrawal parameters versus density. Left: Data for ash, beech and oak, $\alpha = 90°$, 423 values. Right: Data for beech LVL, differentiated by insertion direction, results for $\alpha = 0°$ and 15° deleted, 1732 values.

[128] Even more so, the available range of densities in particular for beech LVL is too small to perform meaningful regression analyses. A broader density range makes any statistical investigations much more robust.

Table 5-5 Results of some test series shown in Figure 5-28 on the left.

Species	N	ρ in kg/m³		F_w in kN	f_w in MPa	
		mean	COV	mean	mean	COV
Screw 1, d_{nom} = 8 mm, L_p = 48 mm, partial thread, normal tip, predrill diameter 6 mm						
Ash	20	645	6.2	8.6	23.0	12.2
Beech	20	690	5.3	10.1	27.1	19.6
Oak	20	674	7.9	7.6	20.2	22.4
Screw 2, d_{nom} = 5 mm, L_p = 40 mm, two threaded parts, special tip, not predrilled						
Beech	12	658	1.2	8.9	46.8	6.1
Oak	11	643	3.0	7.9	41.6	8.2
Screw 3, d_{nom} = 8 mm, L_p = 64 mm, partial thread, special tip, predrill diameter 6 mm						
Ash	20	678	4.7	19.7	38.3	10.7
Beech	20	714	6.7	19.7	38.4	11.6
Oak	20	658	7.4	14.3	27.9	24.2
Screw 4, d_{nom} = 8 mm, L_p = 64 mm, full thread, drill tip, predrill diameter 6 mm						
Beech	10	694	2.2	22.1	42.5	6.9
Oak	10	610	5.7	15.0	28.9	10.2
Screw 5, d_{nom} = 8 mm, L_p = 64 mm, partial thread, normal tip, predrill diameter 6 mm						
Beech	10	694	2.2	21.2	40.8	7.4
Oak	10	610	5.7	13.5	25.9	6.8

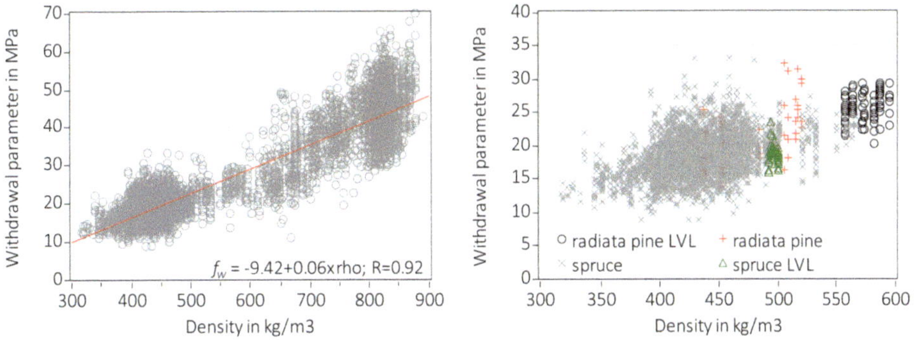

Figure 5-29 Withdrawal parameters versus density for tests at $\alpha \geq 30°$. Left: Excluding results with OSB, particleboard and spotted gum, 5814 values, trendline shown in red. Right: Only softwood differentiated by species, 3659 values.

Further parameters

Generally, if looking at Figure 5-25 to Figure 5-29, the scatter is persistently high, also for homogenised products such as LVL. Therefore, there must be more influencing factors than density or α alone that impact on f_w. The ***tip type*** may further explain the observed scatter[129]. Figure 5-30 however shows that this is not the case; drill tips per se do not lead to lower f_w-values[130]. Consequently, the individual screw properties and natural scatter in timber properties have a strong influence on obtained f_w-values. Recalling Figure 5-28 on the right, such scatter is present even for beech LVL with screws drilled in the face grain, where homogenisation effects are most succinct. Further growth characteristics may contribute, where however, insertion of screws in radial or tangential direction (in solid timber) does not lead to different f_w-values.

Concerning ***predrilling*** as possible influencing factor, both Ringhofer (2017) and Brandner (2019) did not find an influence of predrilling in f_w. Figure 5-31 confirms this in general, where only few data for the largest predrill diameters of $0.8 \cdot d_{nom}$ and $0.85 \cdot d_{nom}$ are available. It is hence difficult to say if the lower values for predrill diameters of $0.8 \cdot d_{nom}$ and $0.85 \cdot d_{nom}$ are significant or not. These large predrill diameters were only used for beech LVL. Looking closer into the results with non-predrilled and predrilled beech LVL, few comparative tests were carried out to investigate the influence of predrilling. The results given in Table 5-6 show that the withdrawal capacity with $\alpha = 90°$ decreases for some of the tested screws when predrilling is applied whereas this effect is less strong at $\alpha = 0°$. However, the tests were not carried out using the same pieces of beech LVL. All results with the largest predrilling diameter of $0.85 \cdot d_{nom}$ stem from tests with screws with two threaded parts. These results are shown in Figure 5-32. Together with Table 5-6, no clear conclusions can be given concerning the effect of predrilling, where it seems that predrilling with large predrill diameters $> 0.8 \cdot d_{nom}$ affects the withdrawal capacity at $\alpha = 90°$ in beech LVL.

Finally, the ***nominal diameter*** d_{nom} must be investigated, which is done in section 5.5.6, after a correction for the density in order to potentially better discern any trends.

129 The development of new tip geometries are usually aimed at reducing the insertion moment.

130 More thorough investigations cannot be carried out as the geometry of the tips is not fully defined, see also footnote 111.

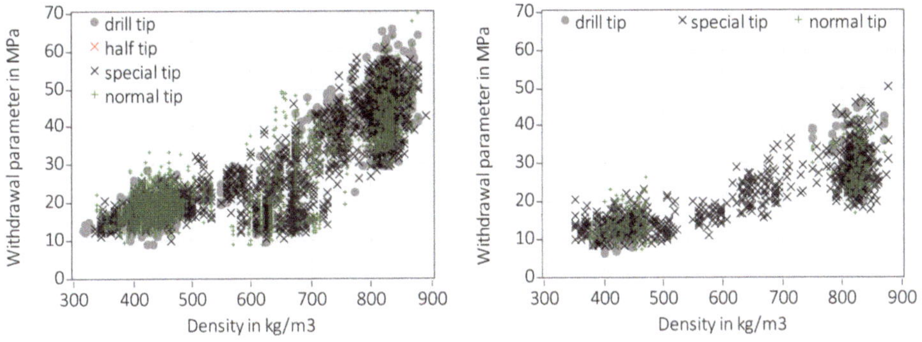

Figure 5-30 Withdrawal parameters versus density differentiated by tip types. Left: $\alpha \geq 30°$, 6398 values. Right: $\alpha \leq 30°$, 1314 values.

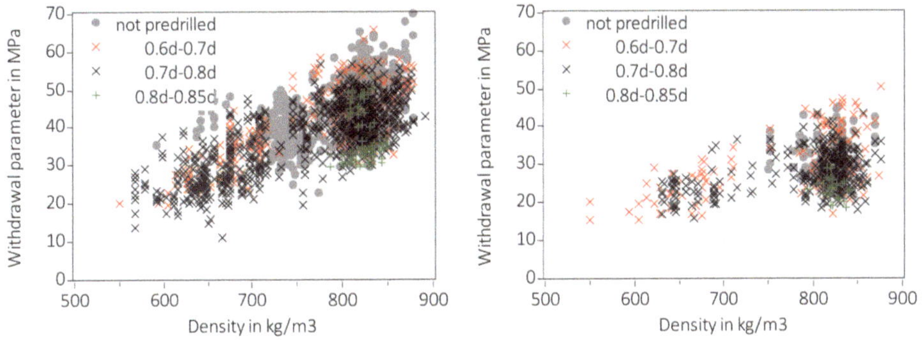

Figure 5-31 Withdrawal parameters of hardwoods versus density differentiated by predrill diameter as percentage of the nominal diameter. Left: $\alpha \geq 30°$, 2155 values. Right: $\alpha \leq 30°$, 701 values.

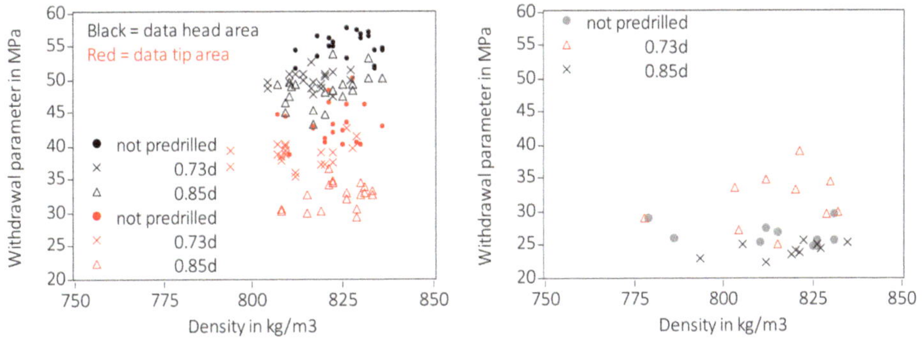

Figure 5-32 Screws with two threaded parts and d_{nom} = 8.2 mm in beech LVL. Left: $\alpha = 90°$. Right: $\alpha = 0°$.

Table 5-6 Results of comparative test series in non-predrilled and predrilled beech LVL.

d_{nom} in mm	L_p in mm	α	Predrill diameter	N	ρ in in kg/m³ mean	COV	min	max	F_w in kN mean	COV	min	max
Fully threaded screw with normal tip												
7.5	45	90°	-	10	820	1.2%	807	836	**17.9**	5.9%	16.6	20.2
			6 mm	10	820	1.2%	807	836	**14.4**	6.2%	13.3	16.2
Partially threaded screw with special tip												
6	36	90°	-	20	834	1.3%	809	858	**11.7**	4.3%	11.0	12.7
			4 mm	20	827	1.7%	804	861	**9.5**	8.3%	8.2	11.0
		0°	-	10	830	1.3%	815	845	**5.9**	11.4%	4.8	6.7
			4 mm	10	843	1.9%	823	875	**5.7**	4.4%	5.3	6.0
8	48	90°	-	21	833	1.3%	809	858	**20.7**	5.9%	18.5	23.4
			6 mm	20	827	1.3%	798	843	**14.8**	6.1%	12.7	16.7
		0°	-	11	824	1.6%	804	853	**11.7**	6.4%	10.9	13.0
			6 mm	10	841	1.3%	828	858	**10.6**	8.4%	9.4	11.9
Screw with two threaded parts and a special tip*												
6.5	39	90°	-	20	822	0.9%	807	836	**11.7**	7.5%	10.5	13.1
			4.5 mm	20	820	1.3%	803	834	**10.2**	3.5%	9.4	10.7
		0°	-	10	814	1.0%	803	828	**7.8**	13.7%	6.3	10.1
			4.5 mm	10	818	1.2%	803	834	**8.4**	12.2%	6.8	10.0

* Thread at tip was withdrawn, thread length was 61 mm. Results for the same screw but with d_{nom} = 8.2 mm are shown in Figure 5-32

5.5.5 Description of observations

Prior to any analysis, all measured parameters must be described and assessed in order to understand their distribution and limitations. Moreover, the relationship between parameters must be assessed; e.g. to check if they are independent from each other and if scattering parameters such as the density show the same variation over the whole bandwidth of test series. Dominant in the database are tests at 90° with non-predrilled spruce and predrilled beech LVL. Table 5-7 gives some descriptive information on these tests, complemented with oak and beech, the other most frequently used hardwood species. Quite an extensive range of nominal diameters was tested and the penetration length differed between series. No distinction was made between solid wood and glued laminated timber. The observed coefficients of variation (COV) of the densities given in Table 5-7 are below 10%, which is the COV of structural timber in accordance with EN 338 (2016)[131]. 42% of all withdrawal tests were carried out with 6 mm and 8 mm screws and a further 26% used 5 mm and 10 mm screws. 46% of all screws had a normal tip, 42% had a special tip, 12% had a drill tip and only 10 tests (one series) were carried out using screws with a half tip.

Table 5-7 Information on withdrawal tests of some species. Missing species and products: Ash and SWP with 40 tests at 90°; spruce LVL, radiata pine and spotted gum with 60 tests at 0°, 15°, 45° and 90°; radiata pine LVL with 120 tests at 15°, 45° and 90°; particleboard with 150 tests at 90° and OSB with 394 tests at 60° and 90°.

	Spruce	Oak	Beech	Beech LVL
N	4012 of which 39 predrilled	331 of which 320 predrilled	172 of which 160 predrilled	2193 of which 1489 predrilled
d_{nom}	3 – 14 mm	4 – 12 mm	4 – 12 mm	3.5 – 13 mm
L_p	$3.5 \cdot d_{nom} - 12 \cdot d_{nom}$	$6 \cdot d_{nom}$ and $8 \cdot d_{nom}$	$6 \cdot d_{nom}$ and $8 \cdot d_{nom}$	$5 \cdot d_{nom} - 8 \cdot d_{nom}$
ρ_{mean} (COV)	434 kg/m³ (7.9%)	661 kg/m³ (6.8%)	690 kg/m³ (5.1%)	816 kg/m³ (3.9%)
$\alpha = 0°$	N = 493	–	–	N = 461
$\alpha = 15°$	N = 20	N = 60	–	N = 100
$\alpha = 30°$	N = 20	N = 60	–	–
$\alpha = 45°$	N = 52	–	–	N = 336
$\alpha = 75°$	–	–	–	N = 100
$\alpha = 90°$	N = 3427	N = 211	N = 172	N = 1196

As usual with descriptive data, numbers such as mean value and COV alone do not fully characterise observations. Two exemplary density profiles are hence further evaluated. Figure 5-33 shows density histograms of the test series with spruce and beech LVL, which constitute 81% of the database, and where additionally, nonparametric (KDE) and parametric density estimations are shown. The histogram for spruce given in Figure 5-33 on the left shows that the density is well represented with a normal distribution. The covered range with a minimum value of 317 kg/m³ and a maximum value of 568 kg/m³ is well representing spruce. The high values between 551 kg/m³ and 568 kg/m³ all stem from one report, in which it is explicitly mentioned that Canadian glued laminated timber with high densities was used; most probably, the used species was black spruce (*Picea mariana*). Figure 5-33 on the right shows the density distribution of beech LVL, where a cluster of lower densities between 671 kg/m³ and 760 kg/m³ can be seen. Due to this cluster, a Weibull distribution

131 For glued laminated timber in accordance with EN 14080 (2013), COV of density ≈ 5%.

fits better to the existing density distribution than a normal distribution. If however this cluster at lower densities would not exist, also here a normal distribution would fit well. A check of the low densities (671 kg/m^3 to 745 kg/m^3) resulted in the observation that they all stem from the first report, in which withdrawal tests with beech LVL are reported[132]. To check this, Figure 5-27 on the left is reconsidered that shows the withdrawal parameter f_w versus the density for beech LVL. Three things can be seen. First, the scatter in withdrawal parameters is high although a highly homogenised product was used; other influences on f_w clearly exist. This leads to the second point, i.e. that the tests with low densities do not lead to significantly lower f_w-values, although they do help to discern a trend of lower f_w-values with lower densities. Such a trend is more difficult to see if only tests with 760 kg/m^3 < ρ < 890 kg/m^3 are considered. Here, the homogenisation effect of beech LVL does not help, because the density variation is small. Third and as already observed, angles α of 0° and 15° lead to lower f_w-values than angle α of 45°, 75° and 90°.

Generally, both histograms show that enough data for the whole density range of the species/product are available. This alone is not sufficient however. Such a comprehensive distribution must range over all independent variables of a regression analysis. Figure 5-34 on the left shows that this is not fully the case for beech LVL, where data for 6 mm and 10 mm screws dominate and where the density range especially for larger diameter screws is smaller. The median values however oscillate around the mean value of 816 kg/m^3. The situation for spruce is more balanced, see Figure 5-34 on the right. All data is combined in Figure 5-35, where the density distribution per nominal diameter (on the left) and penetration length (on the right) ist shown. It can be seen that per nominal diameter, a similar density range was covered, where intermediate densities between 500 kg/m^3 and 700 kg/m^3 are under-represented. This is less the case for the penetration length L_p, where higher densities were not tested with longer L_p. The reason for this is trivial; longer L_p in high-density timbers would have led to tensile failures of the screws. However, a study investigating the influence of L_p on f_w is impossible in this case, because not enough variability of L_p is available. For other parameters such as the pitch p, the density range over the bandwidth of 1.3 mm < p < 8 mm is uniformly distributed, with lowest and highest p only in spruce.

Figure 5-36 shows two further geometrical properties versus f_w. On the left, the ratio of inner divided by outer diameter is shown. The ratio ranges from 0.56 to 0.73 and does not influence f_w, neither for softwood nor for hardwood species. The pitch p of the thread shown on the right ranges from 1.37 mm to 7.99 mm, and also p does not influence f_w.

Finally, the (observed) coefficients of variation are investigated, because considering Table 5-6, it could be postulated that the COV of the withdrawal capacities for angles $\alpha \leq 30°$ are larger than for $\alpha \geq 30°$. This however was not observed by Brander (2019) and Westermayr and van de Kuilen (2020). Figure 5-37 on the left confirms the findings from literature, as

[132] The report dates to 2011; the range of densities of beech LVL was between 671 kg/m^3 and 798 kg/m^3.

there is no difference in COV of F_w for different angles α. The tip type itself does not influence the COV, i.e. there is no relationship of magnitude of COV with a certain tip type. Moreover, the COV is not decreasing for higher densities, i.e. for beech LVL. This was already discussed, because due to the high homogenisation of beech LVL, lower scatter in withdrawal capacities would be expected. Figure 5-37 on the right instead shows the COV of the density per series, where beech LVL has a lower COV.

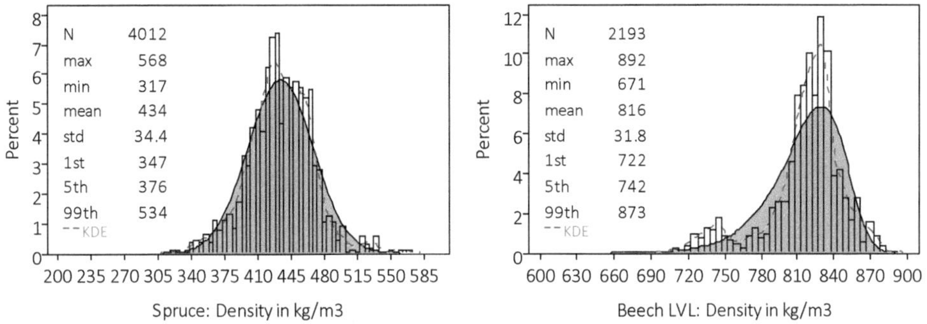

Figure 5-33 Histograms of density with kernel density estimation (KDE) and parametric density estimation. Left: Spruce with fitted normal distribution. Right: Beech LVL with fitted Weibull distribution.

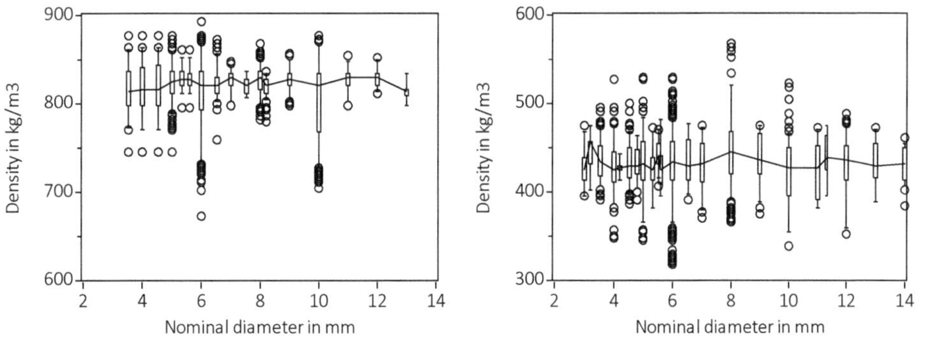

Figure 5-34 Boxplots of density versus nominal diameter (boxplot based on observed values, showing median, interquartile range and 5th percentiles). Left: Beech LVL, 2193 values. Right: Spruce, 4012 values.

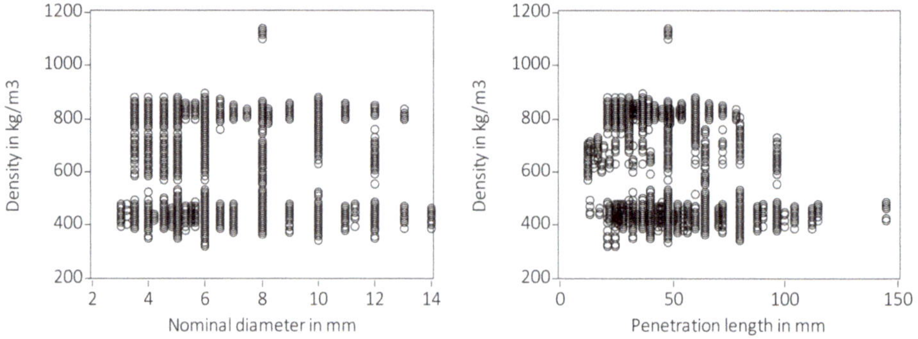

Figure 5-35 Relationship between density and geometrical parameters, 7632 values. Left: Nominal diameter d_{nom}. Right: Penetration length L_{ef}.

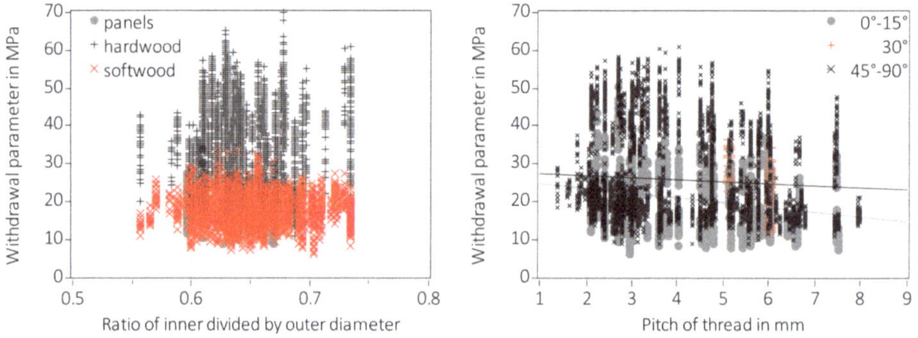

Figure 5-36 Relationship between density and geometrical parameters. Left: Ratio of inner divided by outer (measured) diameter, 7261 values. Right: Pitch of the thread, 6021 values, trendlines for 0° – 15° and 45° – 90° shown.

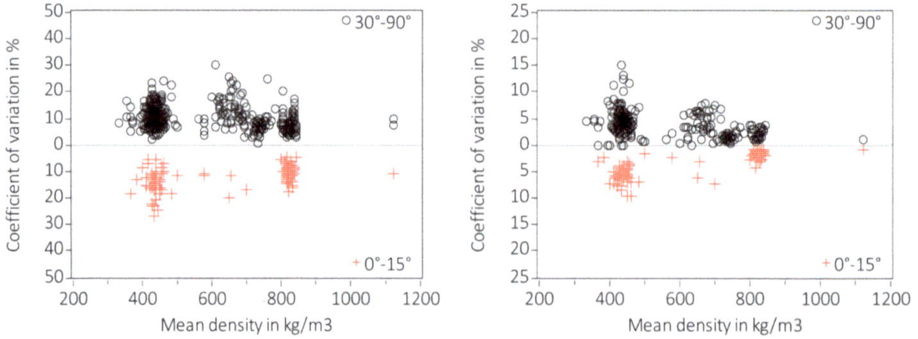

Figure 5-37 Observed coefficients of variation versus mean density per series and differentiated for $\alpha \leq 15°$ and $\alpha \geq 30°$. Left: COV of withdrawal capacity F_w. Right: COV of density ρ.

5.5.6 Relationship with density

As discussed in the previous sections, the withdrawal parameter is mainly influenced by the density, respectively the wood species or timber product, and the angle α between screw axis and fibre direction. Other influencing factors are certainly present, but are difficult to identify in the scatter of test results. Therefore, a correction "eliminating" the influence of density can be carried out in order to get a clearer view on other influencing factors. This is a common procedure, that was already applied in section 4.5.3 to calculate characteristic values ($f_{w,k}$ of nails) and in literature (Brandner et al., 2018). Accordingly, nonlinear regressions in the form of Eq. (5-6) were carried out on the global database and on various subsets.

$$f_w = \alpha \cdot \rho^{\beta} \tag{5-6}$$

where

f_w	Withdrawal parameter in MPa, dependent variable
ρ	Density in kg/m^3, independent variable
α, β	Regression parameters

Eq. (5-6) represents the most widely used and straightforward approach to investigate the dependency of f_w on ρ. Table 5-8 gives results for different subsets ***A*** to ***L***, where the choice of subsets are based on the discussions in section 5.5.4[133]. Obviously, more regressions on subsets, e.g. using only non-predrilled data or subdividing data in groups with certain pitch values, are possible, but are deemed to be a rather diligent but routine piece of work that only leads to more adaptation factors in equations without added value[134]. The residuals were approximately normally distributed and only few studentised residuals had values larger than $|3|$. The coefficient of determination R^2 is with 0.85 largest for subset ***C***, where the most relevant distinction of subsets was made, i.e. the differentiation in angles $\alpha \geq 30°$ and $\alpha \leq 30°$. Furthermore, the "outliers" wood-based panels (low f_w-values at high ρ) and spotted gum (low f_w-values at very high ρ) were excluded. In subset ***C***, the largest amount of individual test data (5774) with the largest variation in test series (440) was available. One of the lowest R^2 of 0.08 belongs to the regression with beech LVL at $\alpha \geq 30°$, which confirms the earlier observation of persistent large scatter in this highly homogenised product[135]. Indeed, no strong dependency of f_w on ρ can be seen in Figure 5-28 on the right. 1632 individual values grouped in 128 test series were available, where most were predrilled (Table 5-7) and some were comparative series using the same screws (see Figure 5-32). This means

[133] N.B.: Category "softwood" includes all softwood species and products, hence also LVL; idem "hardwood".

[134] Additional regression analyses were carried out, e.g. using only predrilled beech LVL at $\alpha = 90°$ (849 individual values), resulting in $R^2 = 0.0095$.

[135] Referring to footnote 134, if only data for predrilled beech LVL and $\alpha = 90°$ are considered, no correlation is observable any more. For more discussion on this, please refer also to section 5.5.7.

that, excluding the comparative series, 57 different screws were used, which may have contributed to the scatter. Also for subset ***H*** "softwood with $\alpha \leq 30°$", a low $R^2 = 0.07$ was obtained, meaning that only a very weak relationship exists between f_w and ρ. Considering beech LVL with $\alpha \leq 30°$, no dependency of f_w on ρ was observed, and a regression using the model in Eq. (5-6) led to no convergence.

As graphical representations are easier to interpret, Figure 5-38 shows the results for some subsets, where the regression curves are drawn over the whole bandwidth of density although they are only valid for certain ranges as indicated by the minimum and maximum values of the density in Table 5-8. On the left and on the right, the trend described by Brandner (2019) and already visualised in Figure 5-26 of a steeper increase of f_w-values with increasing density of hardwoods than of softwoods is again confirmed. In Figure 5-38 on the left showing only regressions for $\alpha \geq 30°$, the influence of only 40 tests with spotted gum (less than 0.02% of all tests) is evident. The regression including spotted gum, subset ***D***, has a very different inclination from the regression excluding spotted gum, subset ***E***. This is trivial, but however, the strong influence is astonishing and the importance of defining clear and correct boundaries cannot be underlined enough. The regression for subset ***H*** "spruce" does not differ from the result for subset ***G*** "softwood" within the valid range of 300 kg/m³ $< \rho <$ 600 kg/m³. This means that a single equation can be used for softwood, at least for the tested products. The results for subset ***F*** "beech LVL" is very similar to subset ***D*** "hardwood including spotted gum". The densities in subset F however ranged only from 671 kg/m³ to 892 kg/m³, making the extrapolations to lower and higher densities questionable. Figure 5-38 on the right shows that a distinction in terms of α is necessary. An approach using one equation for all angles between screw axis and fibre direction overestimates the withdrawal parameters for $\alpha \leq 30°$ significantly, in particular for higher densities, subsets ***B*** and ***I***. Looking at regressions for $\alpha \geq 30°$, one equation could be used for all species however, subset ***C*** compared to subsets ***E*** and ***I***; remembering that softwood species do not have densities larger than 600 kg/m³ and that hence the values at higher densities do not apply. Choosing one single equation obviously punishes e.g. softwood at smaller densities, but manageable design rules may be an important gain for many "everyday" design situations.

Table 5-8 Results nonlinear regression with exponential model according to Eq. (5-6).

Subset	N of tests	N of series	ρ_{min} in kg/m^3	ρ_{max} in kg/m^3	α	β	R^2	N_e^* (min; max)
A All values	7632	579	317	1141	0.0094	1.237	0.65	42 (-4.2; 4.1)
B All excl. wood-based panels and spotted gum	6988	535	317	892	0.0077	1.275	0.73	66 (-3.9; 4.3)
$\alpha \geq 30°$:								
C All excl. wood-based panels and spotted gum	5774	440	317	892	0.0048	1.357	0.85	56 (-4.6; 3.5)
D Hardwood	2155	161	551	1141	0.0170	1.167	0.37	8 (-3.2; 3.4)
E Hardwood excl. spotted gum	2115	159	551	892	0.0009	1.616	0.43	4 (-3.1; 3.2)
F Beech LVL	1632	128	671	892	0.0270	1.102	0.08	4 (-2.8; 3.4)
G Softwood	3659	281	317	595	0.0719	0.915	0.22	25 (-3.3; 5.0)
H Spruce	3499	273	317	568	0.1449	0.799	0.14	22 (-3.3; 5.1)
$\alpha \leq 30°$:								
I All excl. spotted gum	1294	108	353	877	0.0070	1.243	0.73	13 (-2.8; 3.9)
J Hardwood excl. spotted gum⁺	681	55	551	877	0.0304	1.025	0.15	2 (-2.3; 3.4)
K Softwood	613	48	353	595	0.3930	0.581	0.07	2 (-2.1; 3.9)
L Spruce	533	46	353	523	1.1821	0.399	0.02	4 (-2.1; 3.9)

* Number of studentised residuals larger than |3|.

⁺ Beech LVL and oak.

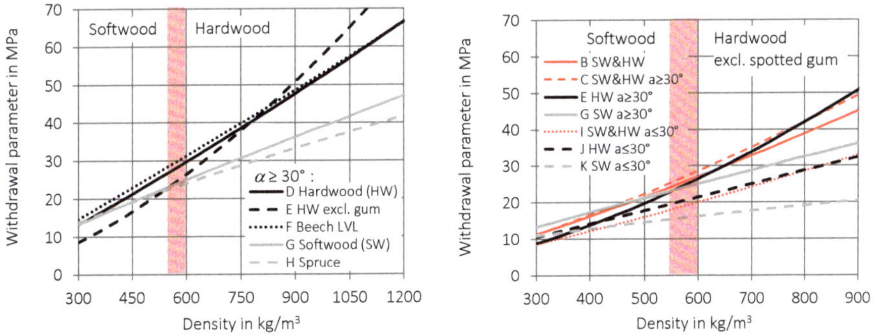

Figure 5-38 Expected withdrawal parameters in accordance with Table 5-8 versus density. The transparent red area indicates the density range where hardwood and softwood overlaps. Left: Results for series with $\alpha \geq 30°$ differentiated by species. Right: Results for all angles differentiated by hardwood (excluding spotted gum) and softwood.

Correction of density

In analogy to section 4.5.3 and Brandner et al. (2018), the regression parameters β given in Table 5-8 can now be used to calculate corrected withdrawal parameters to minimise the influence of density. The dependency of the withdrawal parameters on the density is then reduced and further influencing factors can be better investigated. Therefore, an additional analysis is carried out regarding the influence of α, using corrected f_w-values. The test series using spruce and beech LVL were chosen, as these are over-represented in the database (see Table 5-7). Beech LVL, however, has only a weak relationship between f_w and ρ for $\alpha \geq 30°$, see Figure 5-28 on the right and $R^2 = 0.08$ in Table 5-8, and no relationship was observed for $\alpha \leq 30°$. For spruce, the relationship between f_w and ρ for $\alpha \geq 30°$ is more prominent albeit still weak, see $R^2 = 0.14$ in Table 5-8 and Figure 5-29 on the right. For $\alpha \leq 30°$, the relationship is very weak, see Figure 5-39 on the left and $R^2 = 0.02$ in Table 5-8. The mean density of the 3499 tests with $\alpha \geq 30°$ on spruce was 434 kg/m³ and for the 533 tests with $\alpha \leq 30°$ $\rho_{mean} = 429$ kg/m³. For all 2193 tests with beech LVL, $\rho_{mean} = 814$ kg/m³. Therefore, f_w was corrected as indicated in Eq. (5-7), using the regression parameters β from Table 5-8.

$$f_{w,corr} = \begin{cases} f_w \cdot \left(\dfrac{434}{\rho}\right)^{0.799} & \text{for } \alpha > 30° \text{ in spruce} \\ f_w \cdot \left(\dfrac{429}{\rho}\right)^{0.399} & \text{for } \alpha \leq 30° \text{ in spruce} \\ f_w \cdot \left(\dfrac{814}{\rho}\right)^{1.102} & \text{for all } \alpha \text{ in beech LVL} \end{cases} \tag{5-7}$$

where

$f_{w,corr}$	Corrected withdrawal parameter in MPa
f_w	Withdrawal parameter in MPa
ρ	Density in kg/m³
α	Angle between screw axis and fibre direction

Figure 5-39 on the right shows the results in black for beech LVL and in red for spruce. The differences for different angles are more prominent for beech LVL. Furthermore, the results for beech LVL have a larger variability than for spruce. This large variability may be an indication that the screw itself has a considerable influence; some screws work well in beech LVL, others do not. The difference between using uncorrected instead of corrected withdrawal parameters is low; only outliers differ and the median and interquartile values for angles between 0° and 90°[136]. The lines joining the median values in Figure 5-39 on the right

[136] Only two series in spruce show ±20% difference; where, obviously, one series had the lowest $\rho_{mean} = 332$ kg/m³ and the second series the highest $\rho_{mean} = 561$ kg/m³.

highlight the issue of clustering and resulting necessary interpolations between values. For beech LVL for example, the course of the median line would most probably look different if values for $\alpha = 30°$ would have been available.

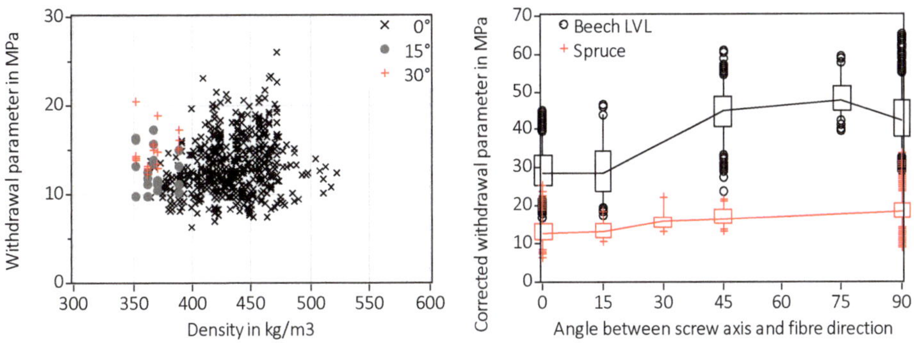

Figure 5-39 Left: Withdrawal parameter for $\alpha \leq 30°$ in spruce, 533 values in 46 series. Right: Withdrawal parameter corrected in accordance with Eq. (5-7); boxplots using 4012 results in 318 series for spruce and 2193 results from 180 series for beech LVL.

Finally, the potential influence of the nominal diameter d_{nom} is investigated. Figure 5-40 shows withdrawal parameters corrected in accordance with Eq. (5-7) versus nominal diameters. On the left, data for the subset *H* "$\alpha \geq 30$ in spruce" is shown and on the left, data for beech LVL. A trend of decreasing f_w-values with increasing nominal diameters can be seen in Figure 5-40 on the left, confirming findings from literature (Brandner, 2019; Frese et al., 2010; Ringhofer, 2017). The reason for this trend may be the same as for nails; i.e. that withdrawal properties depend on other wood characteristics than density alone (Figure 4-26). Indeed, the trend of decreasing f_w-values seems to stop at about $d_{nom} = 6$ mm, which was the upper limit for nail diameters. Similar to the findings for nails, the low f_w-values seem to be constant whereas only the high f_w-values decrease. This observation is less clear for diameters > 11 mm. For these diameters, the number of observations is significantly lower than for the smaller diameters, as can be seen in the number of observations N given in Figure 5-40 on the left. Therefore, it is hard to say how robust the trend of decreasing f_w-values with increasing d_{nom}-values is. For beech LVL, Figure 5-40 on the right, this trend is not present. Also here, less data is available for larger diameters.

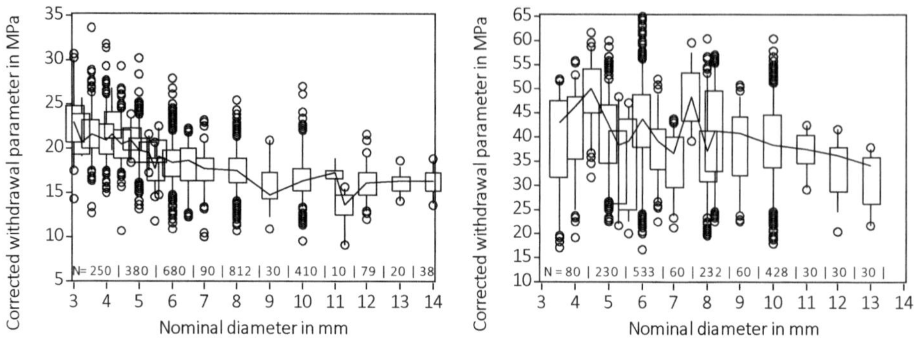

Figure 5-40 Corrected withdrawal parameters versus the nominal diameter. Left: For $\alpha \geq 30$ in spruce. Right: For beech LVL (all angles).

5.5.7 Further statistical investigations

A correction for density alone does hence not lead to a better identification of other influencing factors. Generally, in all analyses carried out up to now, e.g. by creating subsets, no significant reduction of scatter could be obtained which would help to discern trends[137]. At the same time, standard investigations of residuals did not lead to meaningful findings; e.g. in all regressions carried out up to now, residuals were, when investigating their qq-plots (see e.g. Figure 4-7 on the right), approximately normally distributed, their variance was satisfyingly homoscedastic and only few had studentised values larger than |3| (see also Table 5-8). All p-values were "< 0.0001". Although further diagnostic tools available in SAS revealed values with a high leverage or Cook's D, e.g. an exclusion of these values (cut-off if Cook's D > 4/N)[138], no significant changes in regression analyses were observed. Also decile plots did not lead to any identification of misspecified model assumptions (see also Figure 5-41 on the right). Other regression methods did not change the picture either. To illustrate this, robust regression methods are compared to a simple linear regression for one subset. The considered linear regression model is given in Eq. (5-8).

$$f_w = \alpha + \beta \cdot \rho + e \qquad (5\text{-}8)$$

where

f_w	Withdrawal parameter in MPa, dependent variable
ρ	Density in kg/m^3, independent variable
α, β	Regression parameters
e	Residuals

[137] All following tests and analyses were done using both individual and mean values; exemplary outcomes are shown for individual values, Figure 5-41 and Table 5-9.

[138] Such a procedure leads to (ordinary least squares) regression results that correspond to robust regression results using the Huber M estimation with bisquare weighting.

The considered database was subset C of Table 5-8 (all data excl. wood-based panels and spotted gum, $\alpha \geq 30°$, N = 5774). The chosen methods were, as said, a simple linear regression "OLS" using the complete subset (i.e. no values were excluded) and regressions based on "iterated re-weighted least squares (IRLS)". The idea of IRLS is to assign different weights to individual values based on how high the variance of their residual is. With such a procedure, values with large residuals get a small weight and are hence significantly less influential on the overall regression result. With robust regression, the dilemma of standard regression methods, i.e. that outliers should be excluded although there is no compelling reason to do so, can be avoided. The estimators used for IRLS were the Huber M estimation with Huber and bisquare weighting and the more recent MM estimation[139]; both are implemented in SAS. Table 5-9 shows the result of all regressions, where both individual and mean values of f_w and ρ were considered. No significant differences are observed between regressions with individual or mean values or between regressions using different methods.

To underline this statement, Figure 5-41 on the left shows Eq. (5-8) using the values for α and β as given in Table 5-9. To visualise the maximum regression differences, the ratio of results is formed, considering the two methods "linear regression" and "IRLS with MM estimation. A maximum difference of 3% is observed. Concerning the considered simple model as specified in Eq. (5-8) itself, the decile plot in Figure 5-41 on the right shows that this simple model was not misspecified; with the largest deviation in areas with less data as can be seen by the scarce fringe between approx. 500 kg/m^3 and 650 kg/m^3. It must be concluded that more advanced standard regression procedures do not have an added value.

Table 5-9 Regression results based on Eq. (5-8), using both mean and individual values for dependent and independent variable f_w and ρ. Bold values are inserted in Eq. (5-8) and shown in Figure 5-41 on the left

Type of regression	Using mean values			Using individual values		
	α	β	R^2	α	β	R^2
Subset C, Table 5-8	N = 573			N = 5774		
OLS, linear regression ordinary least squares	**-9.562**	**0.065**	0.86	**-9.420**	**0.064**	0.84
IRLS, Huber M-estimation	-9.195	0.064	0.78	-9.002	0.063	0.77
IRLS, bi-square weighting	-8.889	0.063	0.69	-8.740	0.063	0.72
IRLS, MM estimation	**-8.524**	**0.062**	0.49	**-8.407**	**0.062**	0.52

139 For more information on the estimation methods please refer to the SAS handbook.

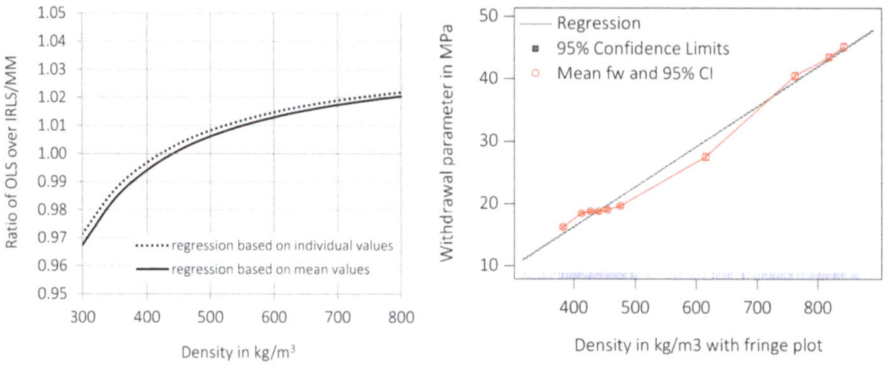

Figure 5-41 Left: Ratio of linear over robust regression results versus density, inserting the bold regression parameters of Table 5-9 in Eq. (5-8). Right: Decile plot of subset C in Table 5-8, with fringe indicating test data density.

5.5.8 Characteristic values

Irrespective of the observed scatter and the resulting challenges in finding meaningful relationships, characteristic values are needed for design. To address this and to understand, how characteristic values scatter, conventional characteristic values are calculated applying EN 14358 (2016) and prEN 14592 (2017) (see also section 4.5.3). The regression results given in Table 5-8 are considered to correct withdrawal parameters in analogy to Eq. (4-16) and to adjust standard deviations in analogy to Eq. (4-17), where however the target COV for the density of beech LVL was 5% instead of 10%, see also Figure 5-37 on the right (and section 5.6.4). In a final step, 5th percentile values for $f_{w,corr}$ were estimated assuming a lognormal distribution and using the corrected standard deviations $std_{fw,corr}$, in accordance with EN 14358 (2016), The limited amount of test results per test series was considered applying the k_s-factor given in EN 14358.

Going to Table 5-8, the regression results for subset C are considered, including tests at $\alpha \geq 30°$ and excluding results for wood-based panels and spotted gum. This leads to a correction factor of 1.36 – the regression parameter β in Table 5-8 – for both withdrawal parameter and standard deviation, and Eqs. (4-16) and (4-17) need to be modified accordingly. For nails, the dependency of f_w on the density was very similar, with a correction factor of 1.35, see Eq. (4-15). However, whereas for nails, only test data with spruce is available, also solid hardwoods and beech LVL were used to determine the withdrawal capacity of screws.

Consequently, different reference densities ρ_{ref} need to be considered when calculating corrected withdrawal parameters $f_{w,corr}$:

$$f_{w,corr} = f_w \cdot \left(\frac{\rho_{ref}}{\rho_{mean}} \right)^{1.36} \tag{5-9}$$

where

$f_{w,corr}$	Corrected individual withdrawal parameter in MPa
f_w	Individual withdrawal parameter in MPa
ρ_{ref}	Mean density of all series per wood type: spruce $\rho_{ref,spruce}$ = 439 kg/m³, solid hardwoods $\rho_{ref,solid\_HW}$ = 670 kg/m³, beech LVL $\rho_{ref,beechLVL}$ = 814 kg/m³
ρ_{mean}	Mean density of test series in kg/m³

The correction carried out in accordance with Eq. (5-9) using one general correction factor is incorrect if different subsets are considered. For instance, the accurate correction factor for subset H in Table 5-8, tests at $\alpha \geq 30°$ in spruce, would be 0.8, and the application of a factor of 1.36 in Eq. (5-9) overestimates the dependency of f_w on the density. To assess the error, the above-described corrections were carried out also for subset H, applying the correction factor of 0.8 and using 420 kg/m³ (= ρ_{mean} of C24, EN 338, 2016) as reference density. The calculated characteristic values are shown in Figure 5-42, where on the left, the values for subset C are shown, applying a correction factor of 1.36, and on the right, the values for subset H are shown, applying a correction factor of 0.8. For the series with maximum and minimum mean densities, obviously, differences between both approaches were highest[140]. The range of $f_{w,corr,k}$-values for spruce is given in the figure caption, and the maximum difference is 5%. Figure 5-42 displays significant scatter, in particular for beech LVL where characteristic values range from 19.4 MPa to 48.3 MPa[141]. Also for spruce, characteristic values differed by a factor of 2, with minimum and maximum values of 7.4 MPa and 16.9 MPa. Figure 5-42 on the left furthermore shows that for spruce and beech LVL, a lower bound (constant) value for the withdrawal parameter could be used. For characteristic values for densities between ca. 550 kg/m³ and 800 kg/m³, a linear increase of the lower bound values with increasing density can be observed, where, however, less data is available. For the subset spruce, a decrease of upper-bound values with increasing nominal diameter may be discerned looking at Figure 5-42 on the right. This statement is not very strong due to the lack of testing data (for larger diameters), but it is similar to what was observed for non-smooth shank nails, see Figure 4-26. An obvious idea is hence to combine databases, which is done

[140] Boundary values of ratio of $f_{w,corr,k}$ for subset H divided by $f_{w,corr,k}$ for subset C were 84% (ρ_{mean} = 332 kg/m³) and 111% (ρ_{mean} = 560 kg/m³).

[141] 19.4 MPa: 6 mm screw at 45° with drill tip, not predrilled, ρ_{mean} = 763 kg/m³; 48.3 MPa: 6 mm screw at 90° with drill tip, not predrilled, ρ_{mean} = 840 kg/m³.

in Figure 5-43. Values for nails tend to be slightly lower than for screws, which is logical seeing the more prominent thread of screws leading to higher withdrawal resistance. Nevertheless, a common lower-bound value seems possible as indicated ***exemplarily*** by Eq. (5-10).

$$(a):\ f_{w,k}=6\cdot\left(\frac{\rho_k}{350}\right)^{0.8}\qquad (b):\ f_{w,k}=6\cdot\left(\frac{\rho_k}{350}\right)^{1.36}\qquad (5\text{-}10)$$

Eq. (5-10) is based on the two exponents $\beta = 0.8$ and $\beta = 1.36$, and on a constant value of 6 MPa for f_w, which corresponds to a withdrawal class in accordance with prEN 14592 (2017, Table 4), see also Figure 4-31, and which is too low for screws, see also Figure 5-42. However, seeing that Eq. (5-10) severly underestimates withdrawal parameters and cannot capture at all fasteners with higher $f_{w,k}$-values, the additional error due to this low constant value of 6 MPa is not deemed to be significant. For design purposes, such a simplified approach is not recommended as with a severe underestimation of f_w, not only capacities are underestimated but also failure modes are not predicted correctly.

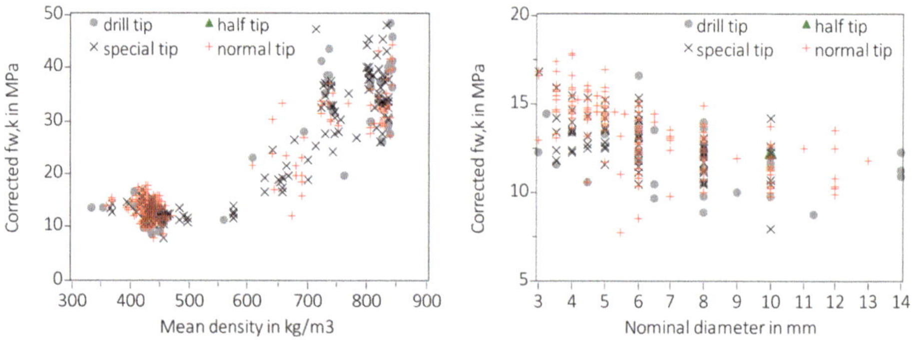

Figure 5-42 Left: Subset C with $\beta = 1.36$ (all tests at $\alpha \geq 30°$ except wood-based panels and spotted gum), characteristic withdrawal parameters $f_{w,corr,k}$ versus mean density, 440 test series; for spruce, $f_{w,corr,k}$ = {7.7 - 17.8} MPa. Right: Subset H with $\beta = 0.8$ (tests at $\alpha \geq 30°$ in spruce), characteristic withdrawal parameters $f_{w,corr,k}$ versus nominal diameter, 273 test series; $f_{w,corr,k}$ = {7.4 – 16.9} MPa.

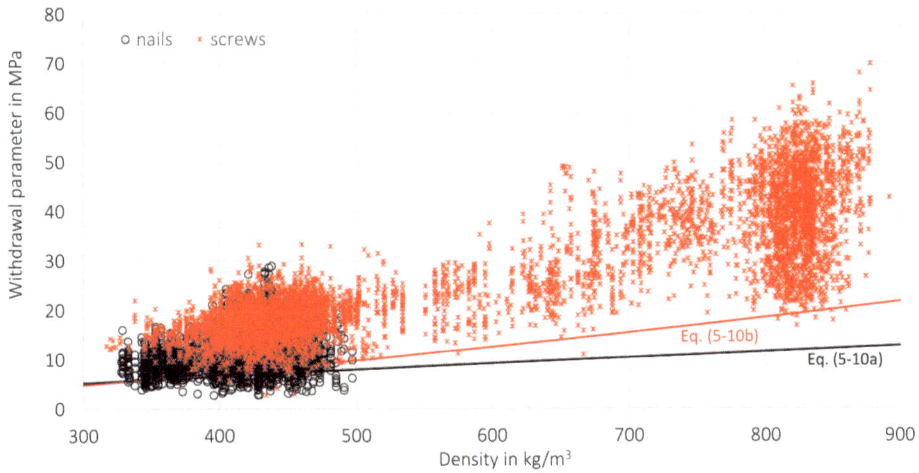

Figure 5-43 Individual f_w-values versus density for nails (including nail tip) and screws.

5.5.9 Conclusions regarding the withdrawal capacity

The withdrawal parameters described and analysed here show a significant scatter, not only when considering the database with 7642 individual tests grouped in 580 series as a whole, but also looking at various subsets. The source of the scatter could not be identified. One contributing factor for this is that not all properties were recorded, e.g. exact tip shapes, thread flank angles or type of coating. These missing properties cannot be considered in analyses that try to identify influencing factors on withdrawal parameters. Moreover, recorded properties concerning the timber encompassed only the global density, and no information on the local wood characteristics directly around the screws is available. If the persisting scatter in test results is considered, there is no scope in defining more and more complex models with more and more adaptation factors that accommodate specific subsets, but cannot accommodate observed variety of screws available on the market. Understanding such scatter puts a focus on the mechanical understanding of withdrawal behaviour and possible optimisations. In terms of design equations, however, it is important to acknowledge that in practice, scatter is present and will certainly be higher than captured in the database.

5.6 Head pull-through capacity

Note: Parts of this chapter were already published in Sandhaas and Blaß (2022).

5.6.1 General

Analogously to the withdrawal capacity, also the head pull-through capacity F_{head} is especially important for joints with screws as screws can transmit high axial forces. However, whereas F_w is always relevant, this is not the case for F_{head}. If the head side member is made of e.g. steel, then – depending on the steel plate thickness – the screw head cannot be pulled through and either steel failure or withdrawal failure from the pointside timber member occurs. Moreover, F_{head} is only relevant for partially threaded screws, which do not have a thread directly underneath the head. Interestingly enough and very different to withdrawal, no publications exist except for few publications covering the head pull-through resistance of nails and screws in plywood and OSB (Chui and Craft, 2002; Munch-Andersen and Sorensen, 2011) and investigations dealing with screw press gluing (Bratulic et al., 2019; Delp et al., 2020; Fürst, 2019), where Delp et al. did not perform head pull-through tests, but measured contact pressure[142].

Currently, characteristic head pull-through parameters $f_{head,k}$ must be taken from technical documentation of screws, hence are proprietary information given as so-called declared values by the individual screw producers. Indeed, the current Eurocode 5 (2010) states that characteristic head pull-through parameters must be determined in accordance with EN 14592 (2012). Concerning head pull-through tests, both EN 14592 and EAD (2016) refer to the test standard EN 1383 (2016). Whereas EN 14592 requires ten head pull-through tests to be carried out, the EAD allows for the declaration of conservative $f_{head,k}$-values without testing[143]. If a screw producer wishes to declare higher values in accordance with the EAD, then twenty pull-through tests must be carried out. A difficulty with EN 1383 is that not enough specifications are given concerning precise test setup and execution. It seems that historically, EN 1383 was drafted for staples, which are pulled-through (thin) wood-based panels[144], as the protocol for staples is rather extensively explained and imprecise specifications concern issues not relevant for these. Particularly the prescriptions concerning the timber thickness t, through which fasteners are pulled through, covers a wide range by stating that "$t \leq 7 \cdot d$". Additionally, EN 1383 only speaks about "the maximum pull-through load F_{max}", which is then used to calculate the head pull-through parameter f_{head}.

142 Aicher et al. (2023) published a paper relevant for the issues presented here, which was not considered in the investigations due to its recent date.

143 For timber products with thicknesses > 20 mm and a characteristic density of 380 kg/m^3, $f_{head,k}$ = 10 MPa.

144 No tests with wood-based panels are contained in the database.

However, the value of F_{max} will very much depend on the thickness through which the fastener is pulled and the deformation at which F_{max} is read. Consequently, it is evident that test results cannot be compared if F_{max} is not taken at the same pull-through deformation.

The test results contained in the database are the maximum values in [kN] reached before or at a deformation of the test machine's crosshead of 15 mm. The thickness of the timber product, through which the screws are pulled, is usually $8 \cdot d_{nom}$, unless noted otherwise[145]. The database contains in total 2854[146] individual test results grouped in 246 series[147], with 10 or 20 tests per series. If steel failures occurred, the test results were not included in the database. The used timber products were mainly spruce (2170 individual tests, or 76% of all tests), followed by beech LVL (11%) oak (8%), beech (4%) and ash (1%). The screws were inserted at an angle of 90° between screw axis and grain direction except for 40 tests at an angle of 60°[148]. All screws in oak, beech and ash solid wood specimens were oriented in radial and tangential direction with respect to the annual rings[149]. This information is not given for all tests with spruce, where 1660 individual tests contain no information concerning the annual ring orientation. Half of the tests per series with beech LVL were inserted in the face grain and the other half in the edge grain. Most screws had a nominal diameter of 8 mm (27%), followed by 10 mm (16%), 6 mm (15%) and 5 mm (10%), with a range of 3 mm to 12 mm.

The used head types covered all four head types given in Table 5-1. Concerning the head type "steel-timber", this is surprising, as usually, such screws are used to fasten steel plates to timber members and hence, no head pull-through tests are carried out. Nevertheless, 180 tests in 17 series with the head type "steel-timber" were carried out (only partially threaded screws). 250 screws had a cylinder head (40 fully threaded screws and 210 screws with two threaded parts, in total 25 series), 810 screws had a washer head (only partially threaded screws, 59 series) and 1624 a countersunk head (90 screws with two threaded parts, 30 fully threaded screws, rest partially threaded screws, in total 145 series). No information about the angle of the countersunk heads was given, except for one series with 60° countersunk heads[150]. Looking randomly at some photos of screws contained in the test reports, it seems that most screws had 90° countersunk heads. Also concerning the finishing

[145] Head pull-through tests through wood-based panels were not carried out. In most reports, the thickness was not mentioned.

[146] For one series with 10 tests, neither the nominal nor the measured head diameter was given. These 10 tests were hence not considered in the analyses.

[147] If also the insertion direction is considered a differentiator, i.e. insertion tangential or radial to the annuals rings or in the edge or face grain of LVL, then the number of series is 327. For one series, the head diameter was not recorded, which is why only 245 series are available for further analyses.

[148] These were tests with screws with two threaded parts inserted in non-predrilled spruce. The results do not differ from those with $\alpha = 90°$.

[149] The orientation of the annual rings has no influence on the head pull-through properties.

[150] Two straight extension lines from the chamfers underneath the head intersect at this angle.

of the side underneath the head, no information is given, where, however, often milling pockets were present. In addition, when checking technical drawings of the screw producers, generally, no information is given concerning the angle of the countersunk head and they only state that milling pockets can be present.

Countersunk heads may need some pre-milling in order to allow for a screw insertion until the head is flush with the surface, especially for beech LVL with its high density. This information, however, is not given in the reports nor can it be retrieved retrospectively. As a rule, it can be said that in a first step, screws were inserted without pre-milling and if it was possible to fully insert these screws without any splitting, no pre-milling was carried out. This procedure is valid for all species and timber products. No judgement can be made concerning any influence of pre-milling, be it partial or full, on the head pull-through behaviour based on the database. The same applies to screws with cylinder heads, where the issue proved less critical due to the small diameter of cylinder heads. Concerning predrilling in general, 2140 spruce, 10 beech and 60 oak specimens were not predrilled. All other test specimens were predrilled before the screws were inserted (i.e. 30 spruce specimens, 100 beech, 170 oak, 20 ash, 334 beech LVL). The predrilling diameters ranged between $0.67 \cdot d_{nom}$ and $0.8 \cdot d_{nom}$, where in 55% of the cases, the predrilling diameter was $0.75 \cdot d_{nom}$.

The used screws were mostly partially threaded screws[151] (87% or 2494 screws), 300 screws (11%) had two threaded parts and fully threaded screws constituted only 2% of all tests (70 screws). A fundamental difference exists between screws with a partial thread and screws with a full thread or two threaded parts. The last two screw types have a thread directly underneath the head that contributes to the head pull-through resistance, whereas partially threaded screws have a smooth shank directly underneath the head. This means that a head pull-through resistance can only be determined for partially threaded screws, whereas for the other screws, it is a "reversed" withdrawal resistance coupled with a screw head that is determined. Screws with a thread directly underneath the head were indeed tested subdividing the series in half of the tests including screw head and thread and the other half was tested pulling through only the threaded part[152], see Figure 5-44 on the left. In the latter case, the given "head diameter" corresponds to the outer thread diameter (= measured nominal diameter). All this data was kept in the database to investigate the effects of a thread underneath the head. If the smooth shank part of partially threaded screws was not long enough to protrude from the timber, i.e. that part of the thread was embedded in the timber contributing to the head pull-through resistance, the timber piece was predrilled in the area of the thread with such a large diameter that the threaded part was loose inside the timber.

---

[151] The threaded part of all screws of the type "TGSHL" in accordance with Table 5-1 was outside the timber and hence, TGSHL screws were combined with the type "TGS".

[152] I.e. the screws were not inserted completely, and the screw head was protruding from the timber. No smooth shank parts were inside the timber in the case of screws with two threaded parts.

The fact that a "classical" head pull-through resistance is determined only for a system of screw head & smooth shank impacts also on the general manufacturing of the test specimens. Partially threaded screws were inserted until the threaded part protruded from the timber specimen, then the specimen including screw was inserted in the testing rig and it was pulled until the screw head was flush with the timber surface[153]; i.e. the upper head side in case of countersunk screws and the lower head side in case of washer head screws. Afterwards, the specimen was unloaded before starting the head pull-through test. Screws with a thread directly underneath the head instead were inserted until the screw head was flush with the timber surface (tests with head and thread), and then the specimen was inserted in the testing rig. Hence, countersunk heads of partially threaded screws were not inserted by drilling them in, but by applying a vertical force.

The ratio of nominal diameter divided by the head diameter was 0.37 to 0.68 for screws with countersunk heads[154], 0.29 to 0.52 for screws with washer heads, 0.72 to 0.83 for screws with cylinder heads and 0.44 to 0.65 for screws with steel-timber heads. The maximum head diameter was 41 mm of a 12 mm screw with a washer head, followed by a head diameter of 29.5 mm of a 10 mm screw with a washer head. Observed coefficients of variation are shown in Figure 5-44 on the right. The coefficients of variation for density are comparable to those observed for the withdrawal test series, see Figure 5-37. The coefficients of variation for F_{head} instead are surprisingly high, with a maximum value of the observed coefficient of variation of 37%[155] and a minimum value of 1.6% with an average of 13.4%.

Figure 5-45 on the left shows the influence of the head types on the head pull-through capacity, and on the right, the results per species are presented. It can be seen that, obviously, for heads with large diameters, i.e. washer heads, higher F_{head}-values on average are reached. Particularly for beech LVL, however, results scatter considerably and screws with countersunk heads reach the highest values. The group of low F_{head}-values between ca. 550 kg/m³ and 750 kg/m³ must be explored further, where it is recalled here that screws with cylinder heads all had a thread underneath the head. As the vertical axis in Figure 5-45 simply shows the measured F_{head}-values in kN, these values must be normalised prior to any further analysis. In general, however, it can already be stated that the scatter observed for all species is surprising, as the head pull-through behaviour is thought to be similar to the behaviour of wood under a compressive load perpendicular to the grain, where the (ductile) compressive strength perpendicular to the grain together with a load distribution area governs and hence, shows rather little variation.

153 As there is no thread inside the timber anymore, the partially threaded screws cannot be drilled in.

154 20 tests in two series were carried out using screws with countersunk heads and a washer. These were assigned to screws with washer heads and the ratio between nominal diameter and diameter of the washer was 0.29 and 0.34.

155 The next largest value was 28%. The series with a COV of 37% was checked and seemed to be correct.

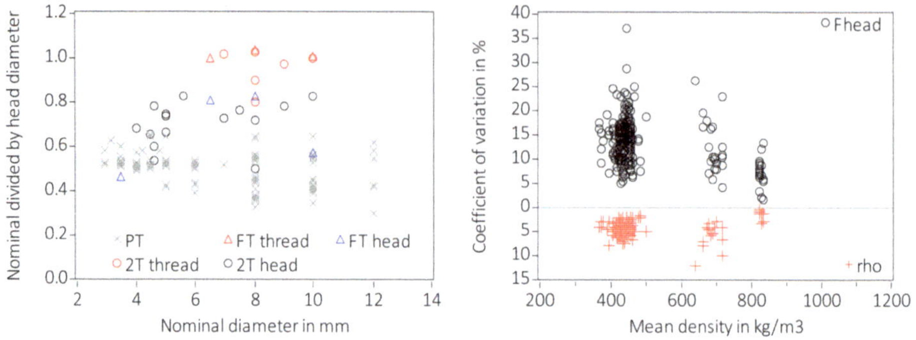

Figure 5-44 Left: Ratio of nominal diameter divided by head diameter. Legend: PT = partially threaded screws, FT thread = fully threaded screws with thread only, FT head = fully threaded screw with head and thread, 2T thread = screws with two threaded parts and with thread only, 2T head = screws with two threaded parts and with head and thread. Right: Observed coefficients of variation.

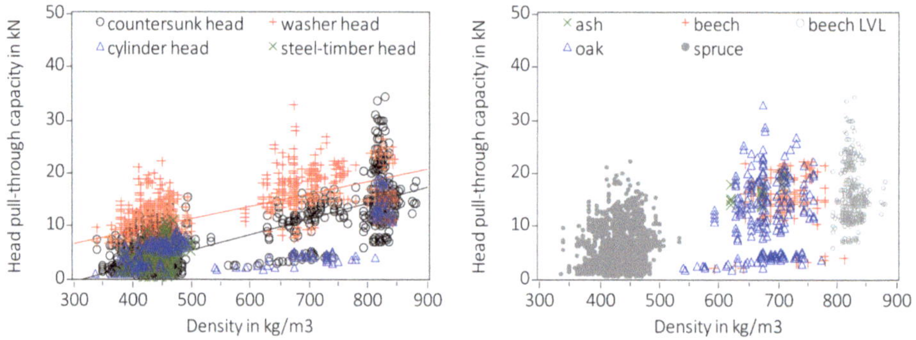

Figure 5-45 Left: Head pull-through capacity F_{head} in kN versus density ρ in kg/m³ differentiated by head types. Trendline is shown for screws with washer heads (red line) and countersunk heads (black line). Right: Head pull-through capacity F_{head} in kN versus density ρ in kg/m³ differentiated by species.

5.6.2 Analysis of head pull-through parameter

Prior to any further comparisons, the head pull-through capacity must hence be normalised. This is done by dividing the capacity through the square of the head diameter, analogously to Eq. (4-18), using the mean measured head diameter. A "head pull-through parameter" for fully threaded screws and screws with two threaded parts, i.e. screws with a thread directly underneath the head, is instead calculated differently and analogously to the withdrawal parameter[156], see Eq. (4-10).

[156] For four series, the thickness t was not recorded, but the respective diameters (3.5, 4, 4.5, 5) were small enough to assume that $t = 8 \cdot d$.

$$f_{head} = \frac{F_{head}}{d_{nom} \cdot t} \tag{5-11}$$

where

f_{head}	"Head pull-through parameter" in MPa of fully threaded screws and screws with two threaded parts
F_{head}	Head pull-through capacity in kN
d_{nom}	Nominal diameter in mm
t	Thickness in mm of the timber piece (≙ penetration length)

Eq. (5-11) was applied for all fully threaded screws and screws with two threaded parts; independently of the test setup, i.e. if head and thread were pulled through or only thread. Consequently and recalling the discussion of section 5.5.2, F_{head} in kN is transformed into f_{head} in MPa using a projected surface, i.e. either through the square of the head diameter or through the product of diameter and thickness.

Figure 5-46 shows the head pull-through parameter versus the density, differentiating between screw types on the left and between head types on the right. Furthermore, the minimum value of 10 MPa of $f_{head,k}$, that can be declared without testing in accordance with the EAD, is shown, see also footnote 143. For spruce, this value of 10 MPa is obviously a good (albeit conservative) choice, whereas a stepwise or linear increase could be introduced for timber products with a density > 500 kg/m³. In general, the scatter is significant, with f_{head}-values for spruce ranging between 9 MPa and 53 MPa and for beech LVL even between 29 MPa and 129 MPa. This scatter is not significantly reduced when showing mean values per series instead of individual testing values, see also Figure 5-50 on the left. Concerning screw types, partially threaded screws can reach higher f_{head}-values; i.e. those screws for which a F_{head}-value can be determined at all[157]. If additionally considering the head type, this seems to hold for partially threaded screws with any head type, whereas for higher densities, particularly for beech LVL, it seems to hold for partially threaded screws with countersunk heads. Seeing the scarcity of data for species with densities > 500 kg/m³, it is difficult to judge if this observed relationship is true or fictitious. However, remembering the previous discussion on the need of pre-milling or not in order to fully insert screws with countersunk heads, it may well be that an insertion of screws with countersunk heads without pre-milling leads to a higher local densification underneath the head resulting in a higher head pull-through resistance. Indeed, if comparing results for countersunk and washer heads in particular for beech LVL, f_{head}-values are lower for screws with washer heads[158], which do not penetrate into the timber pieces (see also Figure 5-48 on the right). However, as no

[157] In general, withdrawal parameters are indeed lower than head pull-through parameters, see Figure 5-43 and Figure 5-59.

[158] In terms of capacity, screws with washer heads will display higher values than screws with countersunk heads.

comparative test series are available and no information on exact head shapes are given, it is difficult to draw reliable conclusions (see also additional tests in section 5.6.3 and Figure 5-50 on the right, where scatter of f_{head}-values for spruce is not different for screws with large or small head diameters).

Comparative test series instead were carried out with solid hardwoods, where the same screws were used in different species. The results given in Table 5-10 show that oak with its lower density has lower f_{head}-values than beech or ash. This confirms the trend of increasing f_{head}-values with increasing density that can be seen for solid hardwoods in e.g. Figure 5-46, and which is not discernible for spruce and beech LVL[159].

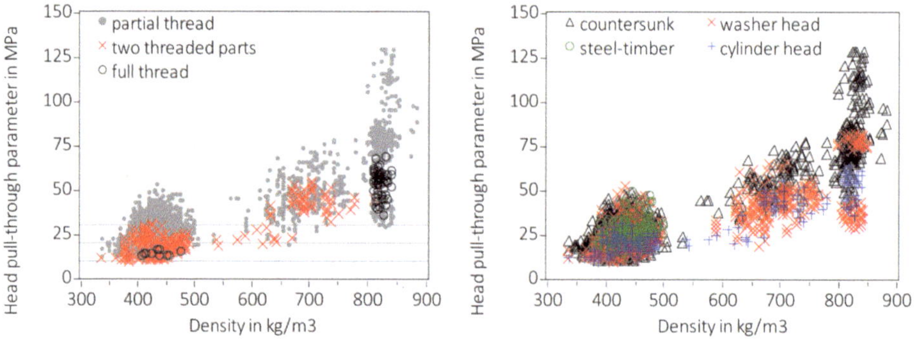

Figure 5-46 Head pull-through parameter f_{head} versus density ρ with horizontal lines at 10, 20 and 30 MPa. Lowest f_{head}-value was 9.3 MPa, all other values > 10 MPa. Lowest value for beech LVL was 29.4 MPa. Left: Differentiated by screw types. Right: Differentiated by head types.

Figure 5-47 shows the results for fully threaded screws and screws with two threaded parts; i.e. the screws where the "withdrawal resistance" was tested. Figure 5-47 on the left shows that for these screws with a thread directly underneath the head, f_{head}-values do not differ, when tests pulling through only the thread (protruding screw head) or the thread and the head (screw head flush with timber surface) are compared. Therefore, the head itself does not contribute to the head pull-through resistance, where the fully threaded screws in beech LVL had both countersunk and cylinder heads without difference in f_{head}-values. Furthermore, screws with two threaded parts followed two different test setups. One standard setup, where the screws were inserted into one timber piece, and a setup, where the screws were inserted through two separate pieces of timber with removal of the timber piece opposite to the head side prior to testing. The purpose was to simulate a realistic insertion

[159] This is a question of the available range of densities. The statement concerning spruce and beech LVL holds also if other axis scales are used or if mean instead of individual values are shown, see e.g. Figure 5-48 on the right and Figure 5-50 on the left.

process, as screws with two threaded parts have different threads at tip and head side. Figure 5-47 on the right shows the respective results. As the procedure of inserting the screws through two pieces of spruce resp. oak (and always including both head and thread) was carried out only in one report, where the standard procedure of using a single piece of timber was not applied, it cannot be said for sure if the observed difference in f_{head}-values is realistic. It is, however, a strong indication that a small comparative study should be carried out to see if this effect of higher values when using a realistic insertion process can be extrapolated to all screws with two threaded parts.

Table 5-10 Test results of series on solid hardwood species. N = number of tests per series, d_h in mm, density ρ in kg/m³, parameter f_{head} in MPa. Coefficient of variation in % is given. All tests with partially threaded screws were predrilled with a diameter of 6 mm.

	N	d_h	Beech		Oak		Ash	
			ρ	f_{head}	ρ	f_{head}	ρ	f_{head}
Screw A[1]	10	6.7	718 | 10%	35.9 | 23%	636 | 12%	25.0 | 26%	-	-
Screw B[2]	20	19.8	714 | 7%	47.9 | 11%	658 | 7%	38.6 | 20%	678 | 5%	44.0 | 12%
Screw C[3]	20	14.8	714 | 5%	65.7 | 12%	684 | 6%	52.9 | 16%		
Screw D[4]	20	21.4	714 | 5%	43.7 | 8%	684 | 6%	39.4 | 10%		
Screw E[5]	20	14.5	714 | 5%	61.6 | 10%	690 | 6%	50.4 | 17%		
Screw F[6]	20	17.5	714 | 5%	49.4 | 4%	690 | 6%	47.7 | 6%		

1 Screw with two threaded parts, not predrilled, d_{nom} = 5 mm, cylinder head, t = 18 mm, head flush with surface
2 Partially threaded screw (PTS), d_{nom} = 8 mm, washer head
3 PTS, d_{nom} = 8 mm, countersunk head
4 PTS, d_{nom} = 8 mm, washer head
5 PTS, d_{nom} = 8 mm, countersunk head
6 PTS, d_{nom} = 8 mm, washer head

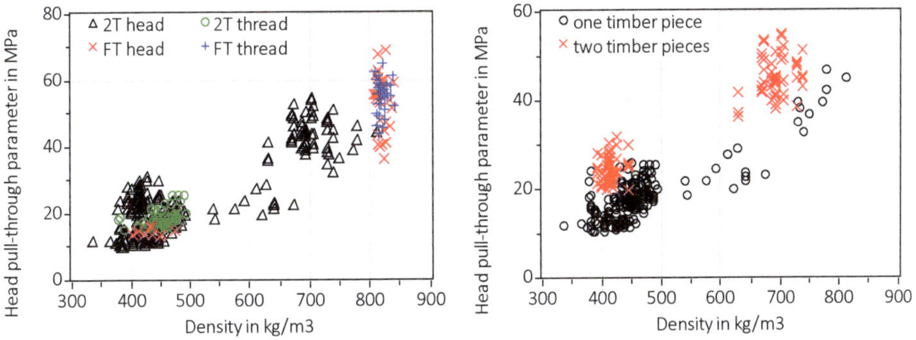

Figure 5-47 Head pull-through parameter f_{head} versus density ρ. Left: Only screws with full thread ('FT') or two threaded parts ('2T'), differentiated by test setup with ('head') or without ('thread') screw head. Screws had only cylinder or countersunk heads. Right: Only screws with two threaded parts and differentiating the two different test setups.

If giving a further look into the tests with beech LVL, further subsets can be identified. The screws in beech LVL were inserted in either the face or the edge grain, where this info is only given for partially threaded screws. However, no difference at all was observed concerning the direction of insertion. If instead differentiating by nominal diameter as shown in Figure 5-48 on the left, a trend of higher f_{head}-values for screws with smaller nominal diameters and hence smaller heads can be observed. Only three reports from 2017, 2018 and 2019 are available for beech LVL, where 220 tests of in total 334 tests stem from the oldest report. If from this report, the results for 5 mm and 6 mm screws differentiated by head types are shown, a clear difference can be observed, see Figure 5-48 on the right. Recalling the discussion on the need of pre-milling of countersunk heads inserted in high-density timber, a possible hypothesis arises. As the screws must be inserted until the screw head is flush with the timber surface, it may be that the countersunk heads of small diameter screws did not need pre-milling in order to ensure proper insertion. This, in turn, may have led to densification underneath the screw head with subsequent higher f_{head}-values.

Alternative calculation of F_{head}

An obvious approach to calculate the head pull-through capacity is to consider its analogy with compression perpendicular to the grain. The most simple approach is to apply a modified version of Eq. (5) given in Leijten (2009):

$$F_c = 3 \cdot f_{c,90} \cdot A_{head} = 3 \cdot f_{c,90} \cdot \pi \cdot \frac{d_h^2}{4} \tag{5-12}$$

where

F_c	Compressive capacity in N
$f_{c,90}$	Compressive strength perpendicular to the grain in MPa
A_{head}	Gross area underneath the screw head

Concerning a value for $f_{c,90}$, the findings of Franke (2008) can be taken into account, who determined compressive strength values for spruce on small cubic specimens (40 x 40 x 40 mm^3) with different annual ring orientation. Franke evaluated a mean compressive strength perpendicular to the grain ***at a strain of 1%*** of 3.7 MPa for tangential compression, 2.6 MPa for radial compression and 1.5 MPa for specimens with annual rings oriented at 45°. Figure 5-49 on the left shows head pull-through capacities contained in the database versus the results of Eq. (5-12) with $f_{c,90}$ = ⅓ · (3.7 + 2.6 + 1.5) = 2.6 MPa[160]. Eq. (5-12) underestimates head pull-through capacities by a factor of three.

---

[160] This mean value of 2.6 MPa is very close to the characteristic value of 2.5 MPa for compressive strength perpendicular to the grain of strength class C24 in accordance with EN 338 (2016).

A second approach to calculate compressive capacities is to assume a stress distribution angle of 45° underneath the head parallel to the grain and of 15° perpendicular to the grain. The usual height (and width) of the timber pieces was $8 \cdot d_{nom}$. Considering an effective height $h_{ef} = 0.4 \cdot h = 0.4 \cdot 8 \cdot d_{nom}$ of the stress distribution in accordance with Eq. (2) given in Leijten et al. (2012) and neglecting the round and not rectangular shape of the screw head, F_{c45} can be calculated as given in Eq. (5-13). As can be seen in Figure 5-49 on the right, F_{c45} is a much better predictor than F_c from Eq. (5-12).

$$F_{c45} = f_{c,90} \cdot A_{ef} = 2.6\,\text{MPa} \cdot A_{ef}$$
$$\text{mit } A_{ef} = \left(2 \cdot 0.4 \cdot 8 \cdot d_{nom} + d_h\right) \cdot \left(2 \cdot 0.4 \cdot 8 \cdot d_{nom} \cdot \tan 15° + d_h\right) \tag{5-13}$$

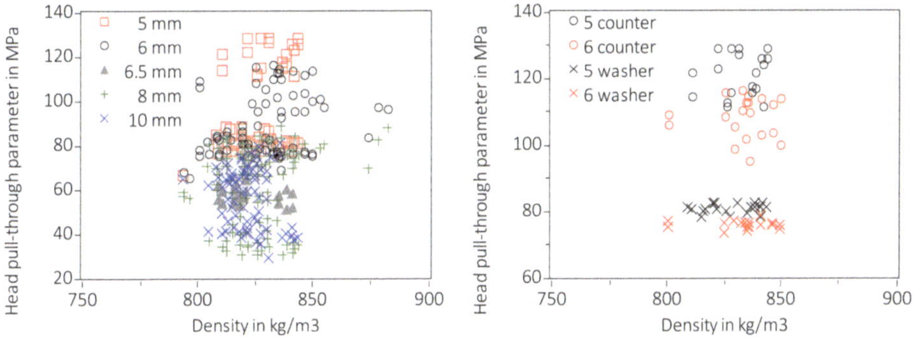

Figure 5-48 Left: Head pull-through parameter f_{head} versus density ρ for beech LVL (all predrilled) and differentiated by nominal diameter. Right: Head pull-through parameter f_{head} versus density ρ. Only results for beech LVL from one report. '5 counter' = 5 mm screws with countersunk heads, '6 counter' = 6 mm screws with countersunk heads, '5 washer' = 5 mm screws with washer heads, '6 washer' = 6 mm screws with washer heads.

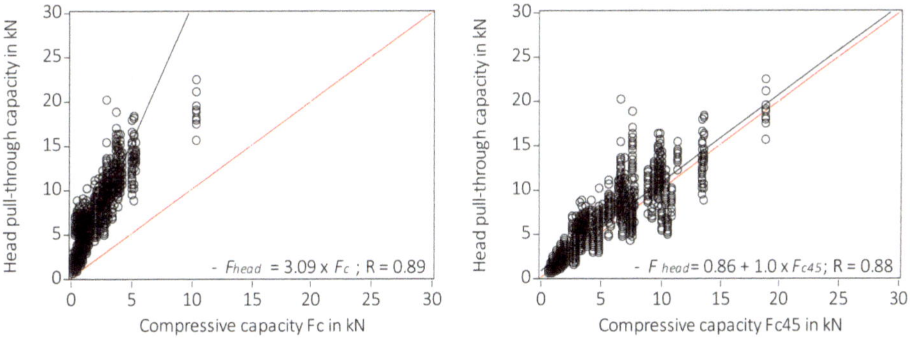

Figure 5-49 Left: Head pull-through capacity versus "compressive capacity" F_c in accordance with Eq. (5-12) with $f_{c,90}$ = 2.6 MPa. The values at F_c > 10 kN are for screws with d_h = 41 mm. Right: Head pull-through capacity versus "compressive capacity" F_{c45} in accordance with Eq. (5-13). Bisect lines are shown in red, trendlines in black. Only data for spruce is considered.

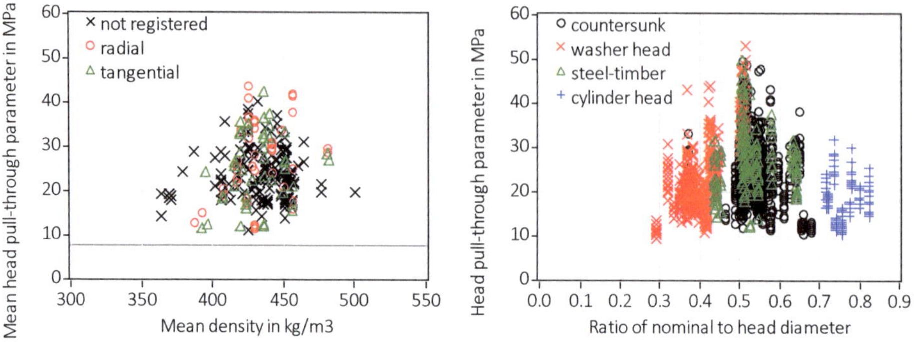

Figure 5-50 Left: Mean head pull-through parameter f_{head} versus mean density ρ for spruce, differentiated by insertion direction. Horizontal line at 7.5 MPa = 3 · 2.5 MPa with 2.5 = $f_{c,90,k}$ for C24. Right: f_{head} versus ratio of nominal divided by head diameter, differentiated by head type. Only results for spruce are shown and tests with thread only were excluded.

To sum up, it is difficult to understand where the variation in test results comes from if it is assumed that a pull-through test is comparable to compressive tests perpendicular to the grain with a small scatter. Differently to compressive tests, splitting underneath a screw head during insertion will influence results. The scatter observed for screws with washer heads, however, cannot be explained with this, at least not fully, because washer heads may not be even on their underside and hence still penetrate the timber. Moreover, it is difficult to control when exactly the insertion process is stopped. Figure 5-50 on the right shows again head pull-through parameters in spruce versus the ratios of nominal to head diameter, excluding the tests where only the thread was pulled through. It is evident that the diameter of the top side of the head in comparison to the nominal diameter cannot explain the observed scatter. Returning to the beginning and the unclear specifications in EN 1383, it must hence be questioned how to interpret the head pull-through capacities, which are read at a crosshead displacement of the testing rig of 15 mm. The following questions arise:

- By how much is a screw head pulled in at 15 mm crosshead displacement?
  During testing, transducers should be used that measure the displacement of the screw head in comparison to the timber surface.
- How is the general load-displacement behaviour?
  Load-displacement curves, using transducers, should be recorded to assess the nonlinear shape of the curve.
- Does the thickness of the timber piece or support conditions influence test results?
  Here, the question is for instance if additional bending of the timber beam influences head pull-through values in cases where a screw is pulled through a rather thin timber member that is supported only at a large distance from the screw.

- Are there any elastic effects?
  Elastic effects will certainly occur, in particular for the tests in beech LVL, seeing that a steel screw with a Young's Modulus of 210,000 MPa is pulled through beech LVL with a certain thickness and a high density. This question is in direct relation to the question before.

- Is there a difference in load-displacement behaviour between screws with countersunk heads and those with washer heads?
  As long as no load-displacement curves are registered, this question cannot be answered. However, there should be a difference, as countersunk heads have a considerable wedge effect on the timber, leading to splitting perpendicular to the grain.

- Does predrilling influence head pull-through values?
  When no predrilling is carried out, more wood material must be pushed aside when inserting a screw, which may influence the behaviour.

- Does pre-milling influence head pull-through results?
  This question is especially important for screws with countersunk heads inserted in high-density timber, where pre-milling may be needed in order to allow for a screw insertion until the head is flush with the surface.

- Is there an influence of the insertion process?
  This question addresses the observation that results for screws with washer heads scatter significantly. For these screws, it is difficult to define a clear end to the insertion process. But also results for screws with countersunk or cylinder heads may depend on how the screws are inserted and what "flush with the timber surface" means.

- Are there geometrical features of countersunk heads that influence head pull-through results?
  This question addresses the observation that results for screws with countersunk heads scatter significantly. As no geometrical data is given for countersunk heads, e.g. the angles of the heads or if milling pockets are present, this observed scatter cannot be assessed. Also different washer heads may lead to different results, e.g. if they have uneven surfaces underneath the head.

- What about head pull-through parameters of inclined screws?
  Screws are often inserted at an angle to the grain, utilising their high axial stiffness and capacity. As the tests assembled in the database are carried out with screws inserted perpendicular to the grain, the effect of different angles on the grain on the head pull-through properties is unknown.

To answer these questions, additional test series must be carried out; with clearly specified boundary conditions and well documented manufacturing and testing procedures. As a consequence, additional small systematic test series presented in the next section were carried out to address some of the above-mentioned questions. These bespoke test series will not be representative for the whole screw population, as such series will be carried out only on very few different screws.

5.6.3 Additional test series

Table 5-11 gives an overview over the additional test series. In all tests, the screw head displacement was measured[161], together with the machine load and machine displacement. Two different test rigs were used (tests with screw A on rig 1, and tests with screws B to F on rig 2), with probably individual influences on the machine displacement. It is this "***machine displacement***" that is currently considered to determine F_{head}-values, and which is used to determine the test results given in Table 5-12, in analogy to all other data contained in the database. The mean moisture content of beech LVL was 6.3% and that of spruce 11.6%. Figure 5-51 shows photos of the underside head shapes of the used screws. In the following, different aspects of the test results are discussed.

Table 5-11 Overview and number of additional head pull-through tests; screws A to C had countersunk heads, screws D to F had washer heads. Width of timber was $10 \cdot d_{nom}$.

	Thickness t in mm	Beech LVL (predrilled 6 mm), m.c. = 6.3%				Spruce, m.c. = 11.6%	
		Not pre-milled		Pre-milled (14 mm)		Not predrilled	
		Edge grain	Face grain	Edge grain	Face grain	Radial	Tangential
Screw A[1]	40	5	5	5	5	10	-
	80	5	5	5	5	10	-
Screw B[2]	60	-	-	-	-	10	10
Screw C[3]	60	-	-	-	-	10	10
Screws D, E, F[4]	60, 80, 100	-	-	-	-	10	10

1 Screw A: Partially threaded screw (PTS), d_{nom} = 8 mm, d_h = 14.8 mm, countersunk head (see Figure 5-51).
2 Screw B: PTS, d_{nom} = 6 mm, d_h = 11.7 mm, countersunk head with milling pockets (see Figure 5-51).
3 Screw C: PTS, d_{nom} = 6 mm, d_h = 11.4 mm, countersunk head (see Figure 5-51).
4 Screws D, E, F: PTS, washer head (screw D see Figure 5-51); Screw D: d_{nom} = 6 mm, d_h = 13.6 mm. Screw E: d_{nom} = 8 mm, d_h = 21.4 mm. Screw F: d_{nom} = 10 mm, d_h = 22.4 mm

161 This displacement was measured not in relation to the timber, but to the "outside"; i.e. the tip of the measuring instrument was touching the screw head and the instrument itself was fastened outside the test rig. A calibrated 20 mm transducer was used for the series with screw A, and for all other series, a special 50 mm transducer was used that could be connected directly to the testing machine, without the need of an additional data acquisition station.

Table 5-12 Test results in terms of mean values with coefficients of variation, thickness *t* in mm, displacement *u* of transducer when machine displacement is 15 mm.

	t	*N*	Variation	ρ in kg/m^3	*u* in mm	f_{head} in MPa	
Screw A, countersunk, d_h = 14.8 mm, d_{nom} = 8 mm							
Beech LVL, pre-milled	40	5	Face grain	835 | 0.5%	10.1 | 9.2%	90.7 | 4.8%	
		5	Edge grain	806 | 1.1%	9.9 | 5,7%	79.2 | 5.9%	
	80	5	Face grain	807 | 0.9%	10.3 | 14.5%	71.2 | 9.3%	
		5	Edge grain	802 | 0.3%	10.4 | 4.5%	82.9 | 2.9%	81.0 | 10.5%
Beech LVL, not pre-milled	40	5	Face grain	835 | 0.5%	12.2 | 2.4%	94.1 | 3.2%	
		5	Edge grain	806 | 1.1%	10.8 | 11.1%	85.5 | 4.0%	
	80	5	Face grain	806 | 1.0%	11.7 | 3.7%	81.9 | 6.9%	
		5	Edge grain	804 | 0.4%	11.4 | 4.1%	82.8 | 6.3%	86.1 | 7.7%
Spruce	40	10		460 | 3.6%	12.9 | 5.8%	24.0 | 9.4%	
	80	10		460 | 3.8%	13.1 | 3.9%	24.4 | 5.8%	24.2 | 7.9%
Screw B, countersunk with milling pockets, d_h = 11.7 mm, d_{nom} = 6 mm							
Spruce	60	10	Radial	466 | 3.7%	13.0 | 4.2%	31.6 | 13.3%	
		10	Tangential	466 | 3.7%	13.0 | 4.5%	28.7 | 11.7%	30.1 | 13.5%
Screw C, countersunk, d_h = 11.4 mm, d_{nom} = 6 mm							
Spruce	60	10	Radial	472 | 3.3%	12.8 | 3.5%	36.2 | 17.8%	
		10	Tangential	472 | 3.3%	13.6 | 3.3%	30.2 | 21.9%	33.2 | 21.6%
Screw D, washer head, d_h = 13.6 mm, d_{nom} = 6 mm							
Spruce	60	10	Radial	469 | 3.0%	11.6 | 4.9%	36.4 | 17.2%	
		10	Tangential	469 | 3.0%	12.4 | 2.2%	29.4 | 10.1%	32.9 | 18.3%
Screw E, washer head, d_h = 21.4 mm, d_{nom} = 8 mm							
Spruce	80	10	Radial	457 | 3.5%	11.1 | 7.4%	23.5 | 7.4%	
		10	Tangential	457 | 3.5%	11.4 | 6.7%	20.8 | 11.2%	22.2 | 11.2%
Screw F, washer head, d_h = 22.4 mm, d_{nom} = 10 mm							
Spruce	100	10	Radial	388 | 3.1%	12.0 | 3.5%	19.6 | 13.3%	
		10	Tangential	388 | 3.1%	12.0 | 2.0%	17.3 | 11.1%	18.4 | 13.9%

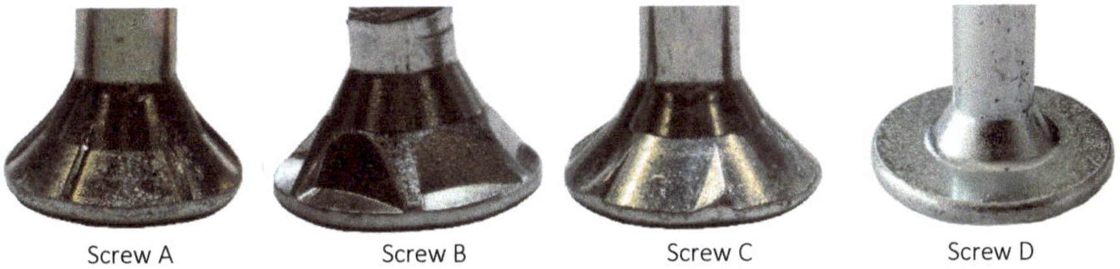

Figure 5-51 Shapes underneath heads (see Table 5-11).

Machine versus transducer displacement

Figure 5-52 shows four load-displacement curves, where the machine displacement and transducer displacement is differentiated. Two systematic differences can be pointed out. Firstly, load-displacement curves for screws with washer heads, upper right figure, show a steady increase until tests are stopped (compression perpendicular to the grain). This is not the case for screws with countersunk heads, which show a more pronounced nonlinear behaviour with a maximum load reached before tests are stopped (wedge effect of countersunk heads). And secondly, whereas the difference in measurement method is small for tests with countersunk heads in spruce, this is not the case for screws with washer heads and in beech LVL, i.e. for tests with more rigid behaviour. It must be underlined that the four chosen curves are by no means representative for all other tests of the same series, and the qualitative load-displacement behaviour may look very different from test to test, see e.g. Figure 5-55. This is, again, astonishing, when considering the analogy of head pull-through tests, above all of screws with washer heads, with compressive tests perpendicular to the grain, which show ductile and qualitatively similar behaviour.

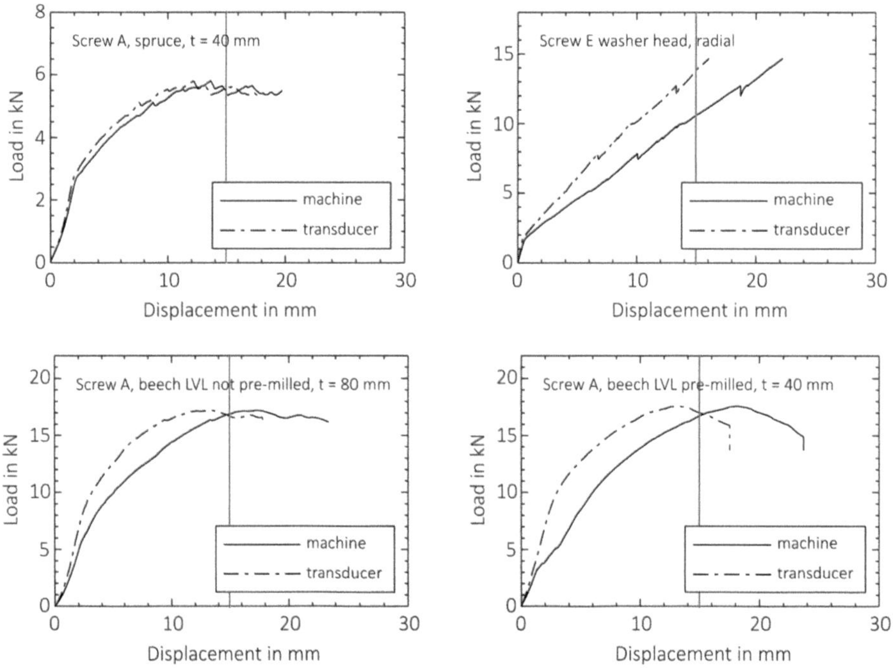

Figure 5-52 Four exemplary load-displacement curves of head pull-through tests. Red vertical line indicates displacement of 15 mm.

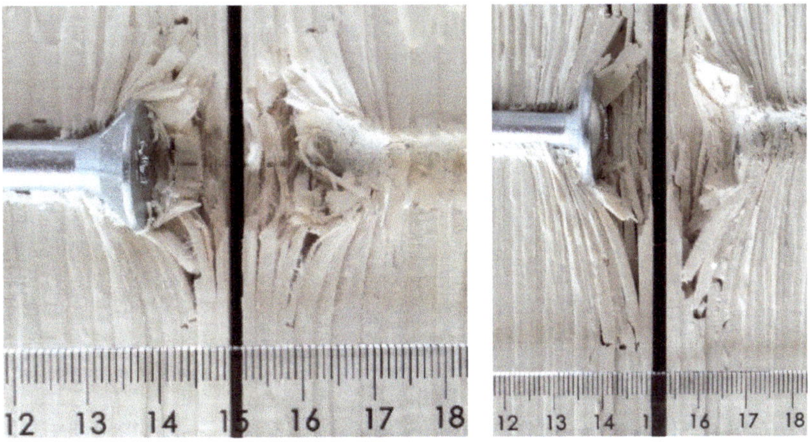

Figure 5-53 Opened test specimens (not predrilled). Left: Screw with countersunk head. Right: Screw with washer head.

Figure 5-53 shows opened test specimens, where the fundamental difference to compressive tests perpendicular to the grain can be seen. In the latter tests, wood fibres are not separated and can act as ropes transferring tensile loads. In head pull-through tests instead, fibres are separated and can be better compared to cantilever beams. However, these simple models ("rope", "cantilever beam") underestimate the influence of shear very considerably as was shown by Bocquet (1997, there Figure 1.22).

In comparison to the tests carried out by Fürst (2019), no initial slip can be observed in the load-displacement curves. The lack of slip facilitates the interpretation as no procedure to determine the curve's origin must be determined. The exemplary curves given in Figure 5-52 show very clearly that the current determination of F_{head} at a crosshead displacement of 15 mm takes place when the load-displacement curves are already far in the ductile range. The permanent deformations visible in Figure 5-53 confirm this; current values of F_{head} go hand in hand with large deformations, and in the case of washer heads, F_{max} is certainly not reached at a crosshead displacement of 15 mm – which represents an arbitrary stop chosen at KIT that was taken in analogy to the specifications given in EN 26891 (1991).

This is an unsatisfying situation, considering the vague prescriptions in EN 1383 (2016) and the effect this may have on results from different testing institutions. Clear prescriptions are needed at which deformation limit head pull-through capacities should be read. These deformation limits may vary depending on the application, similar to the proposal for compression perpendicular to the grain by Windeck and Blaß (2017). Further prescriptions are required as to how deformations shall be measured, in particular for rigid systems, i.e. with high-density timbers and large washer heads, where larger differences between machine and transducer displacement can be observed. Quantitative differences between machine and transducer displacement are given in Table 5-12, where differences of up to

5 mm at a machine displacement of 15 mm were observed. This means, screw heads are pulled in by a maximum of 10 mm at a crosshead displacement of 15 mm. These 10 mm consist of an elastic and a plastic part, as can be seen in Figure 5-53, where the permanent deformation is less than the maximum (machine) displacement of ca. 20 to 25 mm during the test.

Timber thickness

The test specimens were supported such that no bending deformation could occur, i.e. with a steel plate with an opening of 80 x 80 mm^2 to support the tip side timber surface and a distance of 10 to 15 cm between two anchorings fastening the test specimen and steel plate to the test rig. When looking at the results with screw A in Table 5-12, no difference can be observed for the tests with spruce with a thickness of 40 and 80 mm. The same holds for the tests with beech LVL, where, however, the series with 40 mm thick beech LVL and screws inserted in the face grain had higher densities than all other series, obfuscating results.

Pre-milling versus no pre-milling in beech LVL

As expected, no pre-milling to facilitate the insertion of countersunk heads leads to higher f_{head}-values, see Table 5-12 (and Figure 5-54 on the right). However, this trend is only weak, which is confirmed, when looking at the load-displacement curves of all tests with beech LVL given in Figure 5-54 on the left. No difference between the tests with pre-milling and those without pre-milling can be seen. It seems that this difference in production does not lead to the large scatter observed in the database.

Density

Figure 5-54 on the right shows f_{head}-values versus density of the tests with screw A in beech LVL, including the trendlines. The trend based on the in total 40 tests is clear; f_{head}-values are increasing with increasing density. Such a clear trend, however, cannot be observed when looking at the database, see Figure 5-48 on the right. Again, the general issue of discernible trends within bespoke series that, however, vanish when looking at more representative data, can be underlined.

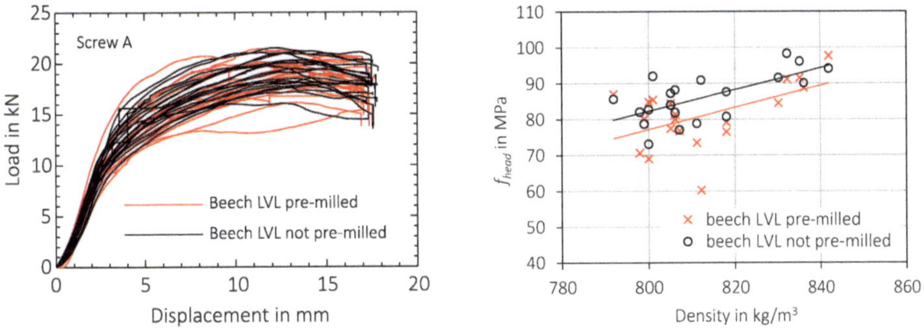

Figure 5-54 Results screw A in beech LVL. Left: Load-displacement curves of head pull-through tests showing transducer displacement. Right: Head pull-through parameters versus density including trendlines.

Insertion direction

Concerning insertion in edge or face grain of beech LVL (screw A), the picture is blurry when looking at Table 5-12, The test results for 40 mm thick beech LVL cannot be interpreted due to the difference in density. The test results for 80 mm thick beech LVL instead show a larger difference between edge and face grain for the pre-milled specimens, and no difference for the not pre-milled specimens (this may be coincidence). Concerning the tests with spruce, tests with screws with countersunk heads (screws B and C) inserted in radial direction lead to higher f_{head}-values, which can be explained with homogenisation effects when inserting screws in radial direction. It is, however, contradictory to the already cited findings from Franke (2008), who evaluated higher compressive strength values in tangential direction in comparison to the radial direction. Figure 5-55 on the left shows the load-displacement curves of screws D, where the largest difference between radial and tangential direction could be observed, which again can be explained with homogenisation effects.

Influence of shapes underneath screw heads

Figure 5-55 on the right shows the load-displacement curves for the tests with screws B and C, which had different shapes underneath their countersunk heads, see Figure 5-51. The scatter is significant, also when looking at the coefficients of variation given in Table 5-12, and this scatter cannot be explained when giving a closer look to the test specimen, production and execution. For screws with washer heads, f_{head}-values seem to decrease with larger screw diameters; a trend that can be confirmed considering the whole database. No conclusive statement can be made based on these few tests, with the exception of the importance of other wood characteristics than the density alone and that different shapes underneath screw heads will impact on f_{head}-values.

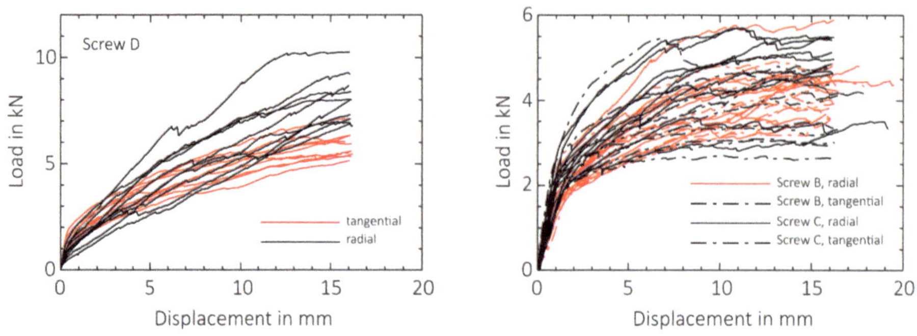

Figure 5-55 Load-displacement curves of head pull-through tests showing transducer displacement. Left: Results for screw D. Right: Results for screws B and C.

5.6.4 Characteristic values

Irrespective of the observed scatter and the resulting challenges in finding meaningful relationships, characteristic values are needed for design. This is not different to the analogous discussion in section 5.5.8 concerning characteristic values for the withdrawal parameter. Therefore, also for the head pull-through parameter, conventional characteristic values are calculated applying EN 14358 (2016) and prEN 14592 (2017). In a first step, a nonlinear regression using all 2854 individual test results was carried out in order to determine a correction factor, the exponent in Eq. (5-14) ($R^2 = 0.7$):

$$f_{head} = 9.5 \cdot 10^{-4} \cdot \rho^{1.67} \quad (5\text{-}14)$$

where

f_{head} Head pull-through parameter in MPa, calculated in accordance with Eq. (4-18)

ρ Density in kg/m^3

This nonlinear regression did not comprise a thorough residual analysis with corresponding deletion of outliers with studentised residuals larger than |3|[162]. Moreover, no differentiation, e.g. with respect to different wood species, was made. Further nonlinear regressions revealed an exponent of 1.003 for all tests with spruce (≙ weak dependency on density, $R^2 = 0.06$), 1.61 for the tests with solid hardwoods ($R^2 = 0.19$), and no convergence was possible for the tests with beech LVL. These results make sense when the limited range of densities available for regression is considered (beech LVL: $\rho_{min} = 794$ kg/m^3, $\rho_{max} = 883$ kg/m^3); the broader this range is, the better are the results which can also be deduced from the R^2-values. A consequential approach is hence to further analyse the data on the "global" scale,

[162] Such an analysis is pointless when scatter, in particular for beech LVL, is high. If beech LVL is considered individually, no relationship between f_{head} and ρ can be found; the range of available densities is too small.

i.e. not differentiating into different wood types. The outcomes and shortcomings of such an approach are discussed in the following.

Therefore, the (global) exponent of 1.67 is used to correct the head pull-through parameters of each test series, hence also of the series with spruce, using reference densities ρ_{ref} that correspond to the mean densities of all series per wood type.

$$f_{head,corr} = f_{head} \cdot \left(\frac{\rho_{ref}}{\rho_{mean}} \right)^{1.67} \tag{5-15}$$

where

$f_{head,corr}$	Corrected individual head pull-through parameter in MPa
f_{head}	Individual head pull-through parameter in MPa, calculated in accordance with Eq. (4-18)
ρ_{ref}	Mean density of all series per wood type: spruce $\rho_{ref,spruce}$ = 433 kg/m³, solid hardwoods $\rho_{ref,solid\_HW}$ = 690 kg/m³, beech LVL $\rho_{ref,beechLVL}$ = 826 kg/m³
ρ_{mean}	Mean density of each test series in kg/m³

The corrected head pull-through parameters $f_{head,corr}$ are assumed to have a lognormal distribution and the logarithm is taken. A normal distribution is assumed for the density. Figure 5-56 shows the histograms of both values for all tests with spruce, confirming the distribution assumptions taken. Using the approach given in Annex D of prEN 14592 (2017), the standard deviations of the corrected head pull-through parameter can now be adjusted so that they reflect the timber population, where a target COV = 0.10 is assumed for spruce and solid hardwoods and a target COV = 0.05 for beech LVL:

$$std_{fhead,corr} = \sqrt{std_{fhead}^2 + 1.67^2 \cdot \left(X^2 - COV_\rho^2 \right)} \tag{5-16}$$

where

$std_{fhead,corr}$	Corrected standard deviation of head pull-through parameter, per test series
std_{fhead}	Observed standard deviation of head pull-through parameter, per test series
1.67	Correction factor, see Eq. (5-15)
X	target COV of density: X = 0.10 for spruce and solid hardwoods, X = 0.05 for beech LVL of timber population
COV_ρ	Observed COV of density, per test series, see Figure 5-44 on the right

Finally, in accordance with EN 14358 (2016), 5th percentile values for $f_{head,corr}$ were estimated assuming a lognormal distribution and using the corrected standard deviations $std_{fhead,corr}$. The limited amount of test results per test series was considered applying the k_s-factor given in EN 14358.

To understand the influence of Eq. (5-16), the ratio of corrected standard deviation $std_{fhead,corr}$ over uncorrected, observed standard deviation std_{fhead} is shown in Figure 5-57. In general, the maximum differences between corrected and uncorrected values were of about a factor of 4, with two exceptions of a 5 times higher corrected standard deviation. Not surprisingly, these two exceptions were observed for the tests with beech LVL whose density may scatter only marginally, leading to high corrections of std_{fhead}. As a matter of principle, the target COV of densities of the timber population was never reached (except for two series, one with oak, observed COV = 12%, one with beech, observed COV = 10%), leading to a correction, i.e. an increase, of the observed standard deviation of the head pull-through parameter. This, in turn, will lead to lower 5th percentiles per series. Indeed, the calculated characteristic $f_{head,corr}$-values based on corrected standard deviations are shown in Figure 5-58 on the left, where in total 14 $f_{head,corr,k}$-values were smaller than 10 MPa (the lower bound value defined in the EAD).[163] This is an effect of the above-described functionalism of Eq. (5-16)[164]. In particular beech LVL with its low scatter of density is punished. The observed COV of the beech LVL used for head pull-through tests is, in fact, 2% (withdrawal tests: 4%, see Figure 5-33 on the left), hence still lower than the assumed 5%. This low scatter of density and its implications was already discussed in Knorz and van de Kuilen (2012), and it underlines the importance of carefully selecting assumed COVs.

Now, $f_{head,k}$-values are re-calculated without considering corrected standard deviations $std_{fhead,corr}$, but using observed standard deviations std_{fhead}. Hence, EN 14358 (2016) was applied, assuming a lognormal distribution, without corrected f_{head}-values and considering k_s-factors, and observed standard deviations were taken without imposing $COV_{fhead} \geq 0.05$[165]. The results are given in Figure 5-58 on the right. In comparison to Figure 5-58 on the left, obtained $f_{head,k}$-values were larger and showed a larger scatter. Reflecting on past discussions on representativeness of data ("Is the scatter of the density available in the database representative enough for practice?") and section 4.4.2 where data was empirically generated to boost databases, the consideration of target COVs and the correction of observed standard deviations analogous to Eq. (5-16) is an important step towards safe values for design. However, a proper choice of target COVs is not straightforward as these data are not readily available for all wood types. Moreover, safe values for design imply an underestimation of capacities, which is not always the design goal.

163 In terms of individual test values, only one original f_{head}-value was smaller than 10 MPa, and the lowest individual f_{head}-value for beech LVL was 29.4 MPa, see Figure 5-46 on the left.

164 Not only, also e.g. the application of k_s leads to small characteristic values.

165 A coefficient of variation of f_{head} smaller than 0.05 was observed in only five of 245 series.

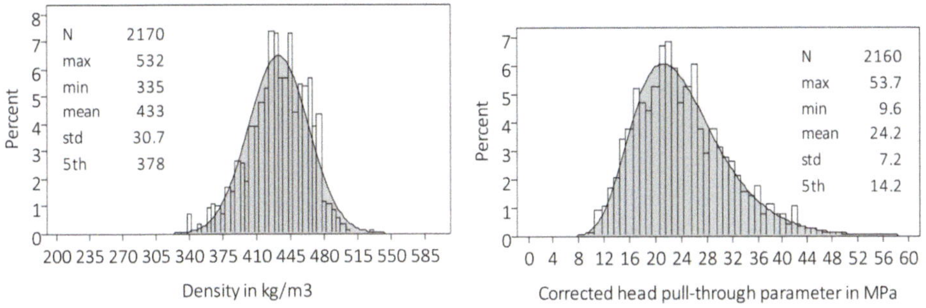

Figure 5-56 Head pull-through tests with spruce. Left: Histogram of density with fitted normal distribution. Right: Histogram of $f_{head,corr}$-values with fitted lognormal distribution.

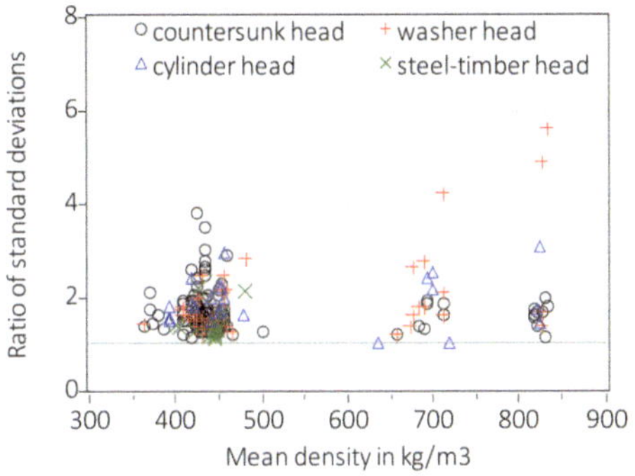

Figure 5-57 Ratio of corrected over uncorrected standard deviations of f_{head} versus mean density per series; corrected standard deviations in accordance to Eq. (5-16).

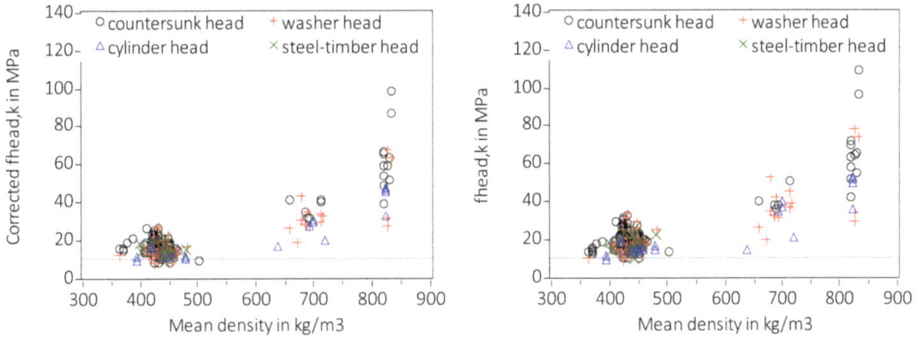

Figure 5-58 Right: Characteristic values of the head pull-through parameter $f_{head,corr,k}$ based on **corrected** standard deviations. 14 of 245 (5.7%) $f_{head,k}$-values < 10 MPa (only spruce).Right: Characteristic values of the head pull-through parameter $f_{head,k}$ based on **uncorrected** standard deviations. 5 of 245 (2.0%) $f_{head,k}$-values < 10 MPa (only spruce).

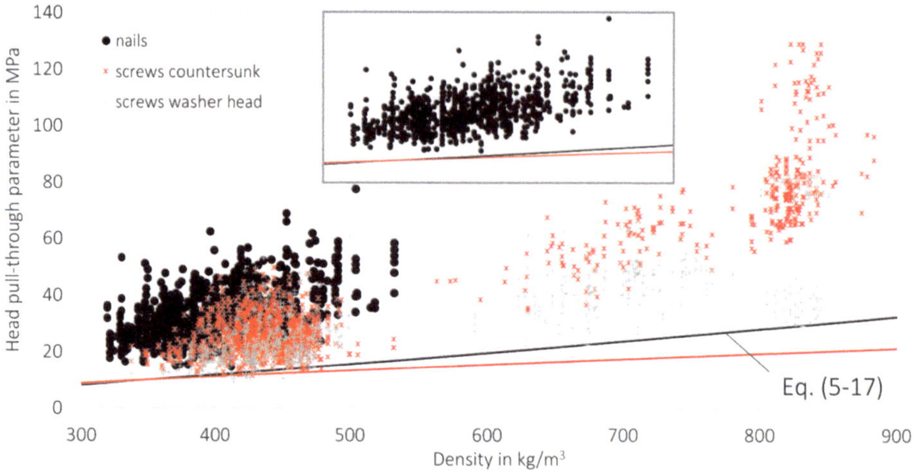

Figure 5-59 Individual f_{head}-values versus density for nails, screws with countersunk heads and screws with washer heads. The box inset shows only data for nails together with Eq. (5-17) for better identification of data cloud position with respect to Eq. (5-17). The red line corresponds to Eq. (5-17), but with an exponent of 0.8 as in the current Eurocode 5 (2010).

Finally, as was done with withdrawal parameters, also head pull-through parameters for nails and screws are combined in Figure 5-59. In section 4.6, a lower bound constant of $f_{head,k}$ = 15 MPa was shown to be a possible scenario for non-smooth shank nails pulled through spruce. The data discussed here confirms the lower bound constant of $f_{head,k}$ = 10 MPa for screws pulled through spruce given in the EAD (2019). This trend of in general lower head pull-through parameters for screws is confirmed when looking at Figure 5-59. Looking at the data clouds for densities of up to 550 kg/m^3 (spruce), the trend of higher head pull-through parameters with higher densities seems to be stronger for nails. This is confirmed when considering the regression results given in Eq. (4-19) for nails with an exponent of 1.33 and the lower exponent of 1.00 found for screws in spruce. However, data for other species than spruce are missing for nails. If it is now postulated that nails pulled through higher density timber behave similar to screws, then the lower bound constants should be adjusted to account for higher densities. If such an adjustment is not made, higher density timber is disproportionally punished. As exponential approaches are very sensitive at their boundaries, the exponent of 1.67 found in Eq. (5-14) should not be considered, as the respective curve increases considerably for higher density timber.

Instead, the exponent of 1.33 found for nails could be applied ***exemplarily***, leading to the following equation shown also in Figure 5-59:

$$f_{head,k} = 10 \cdot \left(\frac{\rho_k}{350} \right)^{1.33} \quad f_{head,k} = 10 \cdot \left(\frac{\rho_k}{350} \right)^{1.33} \tag{5-17}$$

where

$f_{head,k}$ Head pull-through parameter in MPa

10 Constant value of f_{head} of 10 MPa

$\left(\frac{\rho_k}{350} \right)^{1.33}$ Adjustment of f_{head} for density, with reference density of 350 kg/m³ (= ρ_k of C24, EN 338, 2016) and the exponent of 1.33 from Eq. (4-19) for nails

Eq. (5-17) obviously delivers conservative values for the head pull-through parameter. However, the persistent large scatter, in particular for beech LVL, does not allow for any, more sophisticated meaningful regressions. Moreover, considering the load-displacement behaviour shown e.g. in Figure 5-52, remaining conservative may be a good choice as f_{head}-values are currently determined at rather large displacements, and hence represent upper bound values (except for screws with washer heads). Finally, a lack of understanding of the source of the observed scatter, even after having analysed the additional tests discussed in section 5.6.3, makes further analyses rather pointless. It seems that, analogously to what was said previously, also here, other wood characteristics than density alone seems to impact on results. Within a wood type, i.e. spruce, solid hardwood and beech LVL, no or only a weak trend of higher f_{head}-values at higher densities can be observed.

An alternative scenario to Eq. (5-17) could be to keep the lower bound value of 10 MPa for screws pulled through softwood. Analogously, for timber with 500 kg/m³ $< \rho <$ 900 kg/m³, a constant minimum value could be used, without considering any differences in density. Such minimum values for $f_{head,k}$ could be 20 MPa for densities of 500 kg/m³ $< \rho <$ 750 kg/m³, and 30 MPa for 750 kg/m³ $< \rho <$ 900 kg/m³, see Figure 5-46 on the left. This, however, would lead to jumps in equations, which may not be a proper choice.

5.7 Conclusions

A comprehensive database containing test results on self-tapping screws was analysed. Similar to the findings for nails (section 4.7), also all screw properties show a considerable ***scatter***. Also for screws, data about geometrical or material properties is missing as ***this data was not required in the framework of certification testing***, e.g. information on thread flank angle, exact screw tip or head geometry, steel grade and coating is not available. Other missing information that may affect test results concern screw production, e.g. effectivity of hardening or forming of long threads, and the influence of insertion methods, and these may not even be measurable. Consequently, statistical analyses suffer from incomplete data, and the exact sources of the observed scatter could not be identified; solely hypotheses could be developed. Nevertheless, certification tests can provide comprehensive databases where a large variety in timber products, screw types and application cases can be covered. Certification tests lead to representative sampling, since only screws tested in certification tests are on the market and are being used.

Additionally, a serious challenge appeared that affects the development of design equations, which are often based on trends observed in tests. Test series that are available for equation development, however, will be restricted in terms of variability, as not all materials and combinations can be tested, impacting on the representativeness. The database analyses could show that trends that are discernible in bespoke test series may vanish when considering larger and more variable test series. In the case of self-tapping screws, this means that results based on tests with some screws cannot be extrapolated to the whole screw population. It is common agreement that design equations – that are usually based on experimental data or a combination of experimental data and analytical (mechanics-based) models – are valid within certain boundary conditions set by test and/or model limits. These limits should be carefully set, and extrapolations should not be made without further investigations.

Nevertheless, the following concrete consequences can be considered concerning input parameters for joint design in accordance with Eurocode 5 (DIN EN 1995 1-1, 2010):

- ***Steel properties***: The three properties tensile capacity F_t, yield moment M_y and torsional moment capacity M_{tor} were investigated. All three properties show observed coefficients of variation below 5% within test series, but higher coefficients of variation if all data is considered. Consequently, the derivation of general equations valid for the whole screw population comes with the cost of conservative values for certain screws, which will be problematic in e.g. seismic design. Furthermore, the determination of deformation-dependent values for M_y is prone to errors, and more precise descriptions concerning test setup and measuring method should be provided in the respective testing standard.
- ***Insertion properties***: The insertion moment M_{insert} is only needed to ensure that screws do not break during the insertion process. Consequently, the resistance during insertion should be as small as possible. Using the database, however, no optimised screw geometries can be proposed that help keeping insertion moments low. The reason is, as stated above, the lack of relevant data, in particular concerning the exact thread and tip geometry and coating. With regard to practical implications, i.e. that screws must not break during the insertion process, it may be recommendable to use 95th percentiles or even maximum values of M_{insert} to compare with M_{tor}. This is of particular importance seeing the lack of clear regulations concerning the selection of the timber pieces with regard to density. Finally, as long as screws of a certain diameter are produced using the same wire diameter, further optimisation in terms of optimised geometrical relationships in the threaded area are not straightforward.

- ***Withdrawal and head pull-through properties***: Large scatter is observed for both properties, in particular when considering the whole screw population. This is due to the immense variety in screws, but also to wood properties different to density alone. As long as the source of the scatter is not identified, no design equations will be possible that will not punish certain screws.

Contrarily to nails, the definition of technical classes for screws, e.g. technical classes that define characteristic values for the tensile strength or withdrawal parameters, face major challenges. For screws, not only are the steel properties much more complex through the large variety in steel grades, production methods and hardening procedures, but also are input parameters much more connected, because the high axial resistance of screws transforms F_w (and F_{head}) into important design parameters together with the yield moment M_y. For instance, a screw with a high yield moment may possess a low withdrawal capacity, which means that technical classes for M_y and F_w must be connected in order to guarantee correct design.

6 Discussion and conclusions

6.1 General

The main body of this study dealt with analyses of large databases that were retrospectively assembled to address the exemplary seven opportunities formulated in section 1.1. Needless to say, the quality of such analyses depend strongly on the quality of the data itself. Therefore, I would like to open this final chapter citing directly from a publication that does not come from timber engineering, but from information systems (IS) sciences. James R. Marsden and David E. Pingry (2018) write in their abstract:

"We argue that there are major, persistent numerical data quality issues in IS academic research. These issues undermine the ability to replicate our research – a critical element of scientific investigation and analysis. In IS empirical and analytics research articles, the amount of space devoted to the details of data collection, validation, and/or quality pales in comparison to the space devoted to the evaluation and selection of relatively sophisticated model form(s) and estimation technique(s). Yet erudite modeling and estimation can yield no immediate value or be meaningfully replicated without high quality data inputs. [...] As researchers, our empirical research must always address data quality issues and provide the information necessary to determine *What, When, Where, How, Who*, and *Which*."

This is accurately describing the situation also in timber engineering, at least for cases where general issues are discussed and no bespoke models for bespoke issues, all within narrow boundary conditions, are developed. The abstract states the obvious and its last sentence should always be considered, during evaluation of data and in subsequent reporting of – not only experimental, but also modelling – data. Still, recalling discussions on unsatisfying testing procedures or a lack of relevant data, Marsden and Pingry's abstract is relevant. The general data quality in timber engineering is not good (and large) enough to allow for modern modelling and predictive tools based on powerful computing to be applied to general issues. A second aspect comes into play here that complicates matters, and which is the cost-benefit ratio. Fulfilling Marsden and Pingry's demands is no easy task, at least not for timber engineers with their often complex test setups that deliver data, and the assembled certification tests are a perfect example for this. Certification tests lead to representative sampling, since only screws tested in certification tests are on the market and are being used. But these tests are efficiently done and only certain parameters are recorded; any other more exhaustive recording would increase (financial) costs considerably.

Nevertheless, thanks to the large and various databases, different issues could be thoroughly investigated, with all of them addressing one or more of the seven opportunities posed in section 1.1. These encompassed general issues such as investigations on database quality and representativeness of data. But also specific and concrete analyses of sub-issues that arise within the bigger framework of timber joints with dowel-type fasteners could be covered. In this sense, this study allows drawing manifold conclusions and helps to understand implications for research, standardisation or practice.

6.2 Consequences for research

The two extensive databases available for this study covered certification test results for nails and self-tapping screws, containing many input parameters such as yield moments, tensile capacities or withdrawal and head pull-through capacities necessary for design of timber joints. In both databases and for all investigated parameters, a large and persistent scatter was observed. This was true also for the small database assembled for staples. In all data analysed up to now, bespoke series may deliver clear trends. If, however, representative data is considered, these clear trends vanish, and sources for the scatter cannot be identified. The most probable reason for this is the much larger variability of the databases in terms of materials (both fasteners and timber products) than generally used in tailor-made test series, together with some inherent information gaps of the databases, e.g. missing information on geometrical features of the fasteners that must not be recorded in the framework of certification tests. Tailor-made series, of course, are an important tool for mechanical understanding and potential optimisation, where in general, well-defined materials are used that set the boundary conditions for any derived models based on mechanical, numerical or statistical approaches. Such boundary conditions are for instance a range of densities or fastener diameters. Results are often extrapolated to all e.g. self-tapping screws that fall within the boundary conditions. The databases show that such extrapolations are questionable. This is not all though; also issues difficult to record impact on test results. In particular, human influence relating to specimen preparation and test execution, the influence of fastener production as well as the influence of the timber material directly around the fastener are difficult to record. Moreover, research efforts put into understanding sources of scatter should always acknowledge that in practice, scatter will certainly be higher than captured in any database; the additional value of knowing e.g. "wood properties directly around the fastener" may be low as fastener insertion on the building site is random.

All these issues impacting on observed scatter are amplified when results from different testing institutions are considered. Often, databases are considered as universal remedy, and this study could show that this opinion does not hold, albeit not without being aware of database limitations. This situation is exacerbated when considering other tests such as embedment tests or, above all, joint tests, which are subject to even more challenges with respect to database assembly due to more complex test setups and influencing factors. A natural consequence for any future test series is a proper and thorough data collection during the whole test campaign (N.B.: tests in the framework of research, not certification). This includes measurement of all material data, precise documentation of the test setup and execution including information on force and displacement measurements. Concerning material data, all data should be determined also in cases where this means enlargement of test campaigns. If joint tests are carried out for instance, all input parameters should be determined as well, i.e. fastener, embedment and withdrawal properties. All tests where deformations play an important role must be documented and analysed properly including clear statements at which deformations capacities are read and how failure modes developed. Timber densities and moisture contents should be always measured, at relevant locations, i.e. close to fasteners. Concerning processing of test results, boundary conditions must be strictly kept in mind, including seemingly irrelevant conditions. Indeed, this study could show that the common understanding of low variation in steel properties of (timber) fasteners is wrong, if the whole fastener population is considered. Due to uncertainties related to production, post-production treatment and their effects on individual batches, steel properties potentially scatter as much as timber properties. The situation is different from steel engineering, where in general standard bolts are used, and where geometric and production requirements are strict.

Besides boundary conditions, also database bias must be kept in mind, i.e. uneven distribution of available test data for further analyses. Moreover, no globally valid regression equations can be derived using limited datasets, and regressions that hold for bespoke test series may not be valid anymore if a whole population is considered. The fulfilment of above-listed full documentation should become a critical element during e.g. the review process of publications, which, by the way, is not difficult to fulfil as it does encompass merely diligent documentation of test preparation, execution and results.

An obvious answer to the discussed issues with data quality is to carry out more experimental tests, e.g. to close database bias or record better geometrical data. This is, however, not expected to alleviate the situation as scatter is expected to be persistent, because many influencing factors can simply not be measured, e.g. human influence. This is anticipated to hold even if documentation is improved, as there are too many influencing factors and as overall variety, particularly of screws, is enormous. Modelling work instead can be intensified, e.g. by generating simulated data to boost databases (instead of gathering test values) and by including stochastic approaches in (numerical) modelling, but not by applying standard statistical regression tools to derive yet another set of regression equations, adding yet another factor to simulate better precision. Additionally, other disciplines may provide interesting perspectives for timber engineering, see section 6.5.

6.3 Consequences for standardisation

6.3.1 Design standards

The observed large scatter has direct consequences for design standards as long as the source of the scatter cannot be identified, which would potentially result in design equations with precise boundary conditions (and/or optimisation of fasteners). Only then, design equations can reliably predict the modelled parameters without being overly conservative. At the current state, different scenarios exist how design standards can deal with the observed persistent scatter:

- A system similar to steel engineering is chosen, where fasteners are standardised, in terms of both materials and geometry.
- Design equations or constant lower bound values are included in the design standards, accepting that design is conservative. It must be underlined that this potentially leads to wrong predictions of the failure modes.
- Technical classes are defined that allow for design equations with different constants whilst better representing the database (see also section 6.4).
- Design standards refer to technical documentation where input parameters are declared. This means that only the design models, e.g. the *EYM*, are contained in the design standard, but necessary input parameters must be taken from proprietary producer information.

The choice of any scenario must be taken by code writers, practitioners and fastener producers together. Possibly, different approaches can be chosen, where simple and conservative solutions could cover, say, 80% of all timber design cases, whereas more complex solutions are required for highly engineered timber structures. "Simple and conservative solutions" could correspond to the second bullet point that describes the current format of Eurocode 5 and which seems outdated considering that in every engineering office, computers facilitate at least the programming of spreadsheets for design, that are able to accommodate e.g. a large variety of input parameters. Additionally, commercial software packages can be extended to cover a certain amount of scatter, e.g. contain different screw types with different properties. Recalling the uncertainties also on the side of the actions on structures, however, it must be questioned how outdated current procedures are and if these simple equations or values, e.g. one for the yield moment and one for the withdrawal capacity, are not sufficient for many design cases. The "more complex solutions" then would encompass for instance numerical modelling approaches (e.g. *BOF*-models as presented in Schweigler et al. (2021)) or advanced analytical models endorsed by testing. This is possible already today, e.g. looking at the example of the Louis Vuitton-building ("Fondation Louis Vuitton pour la Création") in Paris (Bocquet et al., 2012).

6.3.2 Test and product standards

Imprecise test and product standards are certainly a cause for the observed scatter and cast doubt on test data and their comparability, in particular if data from different institutions is used. Three main sections of such standards should hence be improved; (i) assembly and properties of materials (fastener, timber product), (ii) precise description of test setup and execution, (iii) precise information on measurement and evaluation. This is no easy task, as an equilibrium must be found between minimum reporting in the framework of certification testing and necessary reporting in the framework of research. All this depends also on the scenarios outlined in the previous section, as these will impact on certification tests. A probably remaining issue will be the comparability between different testing institutions. For this, the idea of "reference fasteners" formulated by Munch-Andersen and Svensson (2016) could be considered and further developed, including round robin tests – which are already required by accreditation agencies by the way. The significance of this idea must be thoroughly checked, as such reference tests involve a lot of resources, and the same material must be delivered to different testing bodies. The wording "same material" exemplifies the challenges. Fasteners must be the same, which in the case of ringed shank nails means nails from the same batch and produced at the same time to avoid any wear of the production machines that could lead to less sharp rings. Also timber material must be the same, which is basically impossible due to the natural scatter. In any case, a campaign of round robin tests should be used to improve both product and testing standards (for certification), as they will help to highlight missing specifications and to shape standards in an unequivocal way.

6.4 Consequences for practice

This section is directly related to the possible scenarios outlined in section 6.3.1. Either the need to consult technical documentation will remain or practitioners will be provided with more or less conservative values. Independently of any decision taken, they should be sensitised to the impact of scatter on predictions of stiffness and capacity. The introduction of technical classes is widely discussed as a viable means to tackle scatter whilst catering for simple rules. For nails, it is possible to allow for different steel properties, whereas in the case of withdrawal and head pull-through, production-related issues and wood characteristics other than density impact on results and lead to a high scatter that is difficult to capture reliably. In the case of screws, the introduction of technical classes is challenging and not possible in a straightforward manner. This is due to what was already outlined in section 5.7. For screws, the steel properties are much more complex through the large variety in steel grades, production methods and hardening procedures in comparison to nails and staples. Additionally, analogous to nails, the steel properties depend on the diameter of the screws, which must hence be considered as well. This is not all; the input parameters for design are much more connected than for nails and staples, because the high axial resistance of screws transforms above all the withdrawal capacity into an important design parameter together with the yield moment. This means that technical classes for these properties must be connected to each other, although they are not related. Finally, it is important to remember that capacities are needed for design, and not parameters, meaning that an 8 mm screw inserted 100 mm in the timber may have the same withdrawal capacity than a 10 mm screw inserted 80 mm. Therefore, focussing on parameters instead of capacities may not lead to pragmatic solutions either.

Recalling the current system of a mixture of values given in design standards and values given in Declarations of Performance, opinions differ as to whether this system is practice-oriented or not. European Technical Assessments allow for a high speed of innovation whilst forcing design engineers to consult a myriad of proprietary and regularly updated technical documents and to perform building site inspections that guarantee the use of appropriate fasteners. Recalling what was said in section 6.3.1, perhaps a simple system based on standardised fasteners optimised for structurally simple timber buildings up to three storeys would alleviate this situation. Nonetheless, highly engineered, innovative fasteners would still be available for high-performance timber structures designed by expert engineers. As a consequence, a parallel system of simple rules and values for 80% of the design cases and rules based on technical documentation would continue to exist.

6.5 Perspectives

So how can this situation be improved? On the one hand, the sources of scatter need to be identified, which will result in a better understanding of the mechanical behaviour with subsequent improved modelling, designing and optimising, and on the other hand the observed scatter needs to be incorporated into model and design approaches, or, at least, information on lower and upper boundaries, as in practice, scatter will certainly be even higher. Approaches from other disciplines may help, where not all are new, but merit a new try thanks to the considerably increased computer power available nowadays. A fundamental problem here is that the assembled databases, despite the in parts high number of test values, are by no means "big data". Here, databases are limited and imperfect. Advanced statistical methods to deal with incomplete datasets exist (Radhakrishna Rao et al., 2008). Concerning censored data, such approaches were already applied to timber engineering (Steiger and Köhler, 2005).

Concerning better understanding of the mechanical behaviour, it is crucial to exclude influencing factors that cannot be quantified. This covers for example human influence on production and assembly of joints. For this, methods used for product development of power tools could be applied to exclude human influence during test specimen manufacturing (Uhl et al., 2019). It may be possible that, once human influence is excluded by using test benches, test results of e.g. withdrawn screws or nailed joints loaded in tension can be better interpreted. Using measurements from power tools to predict fastener properties are already investigated by Eckert et al. (2023), who found a strong correlation of the energy needed to insert a self-tapping screw with the screw's withdrawal resistance.

Concerning incorporation of scatter into modelling, the databases assembled in this study, incomplete as they are, pave the way for advanced numerical modelling by significantly extending available input parameters including information on their statistical distributions. Modern approaches in computational mechanics involve for example so-called data-driven formulations (Eggersmann et al., 2019), where no classical constitutive material models are applied, replacing the description of material behaviour with statistics and self-learning algorithms using direct experimental data instead. In a broader sense, these approaches belong to data science research that applies statistical tools and machine learning to large datasets in order to make predictions of the output (Baesens, 2014; Liu et al., 2019). Indeed, current computer power could be used to re-look into neural network model approaches (Blaß and Gard, 1994). Still, it must be kept in mind that there will be no progress in knowledge if only the analyses are getting more complex, based on incomplete datasets, without comprehension and validation of the physical basis.

6.6 Closure

This study started with seven exemplary opportunities associated with database assembly, which were addressed in various parts of this study and to which I would like to give short and concise answers. ***Opportunity 1*** asked if better design models can be derived if more comprehensive test data is available. The answer is no; not as long as influencing factors on test results cannot be identified. It is possible to derive design models, but they will be conservative due to the large scatter, and any refinements of models are window dressing. ***Opportunity 2*** stated that databases allow for validation of existing rules and of extrapolations. This statement holds; but, again, the large scatter comes into play as it may obfuscate trends. The assembled databases could underline how crucial correct boundary conditions are and that no extrapolations should be done without proper investigations and based on only small tailor-made test series. ***Opportunity 3*** postulated that issues can be revealed that may not have gotten the attention they deserve. Yes, this is the case. Such issues encompass the need of careful and exhaustive testing documentation, importance of representativeness of data when extrapolating results and large scatter when considering representative data instead of few series. Concerning testing documentation, the cost-benefit ratio addressed in section 6.1 comes into play; testing in the framework of certification is different to testing in the framework of research. The answer to ***opportunity 4*** is obvious; of course database assembly helps to critically reflect past work. In particular, good data quality is of major importance not only for deterministic but also for predictive models. This last sentence directly leads to the last three challenges, as these are difficult to tackle due to a lack of good enough data. A first step, tackling ***opportunity 5***, would be to apply advanced statistics, e.g. statistics for incomplete datasets, to investigate what else we could learn from the assembled databases. This should be done in collaboration with experts from statistics. Only after having been able to further differentiate scatter into "controllable scatter", i.e. properties that depend on defined features such hardening procedures, and "non-controllable scatter", i.e. human influence affecting borehole quality, the architecture of any extended software can be designed, ***opportunity 6***. Whereas the first type of scatter will define how many different fastener types in combination with timber products will be included as options in design software, the latter type of scatter will define percentile values and safety margins. ***Opportunity 7*** finally addresses, maybe, how design could be carried out in the future, leaning on artificial intelligence in the broadest sense. For years to come this will not be easy to achieve; seeing the complex nature of even a simple property as the tensile strength of a timber screw or the difficulty in judging if results are physically sound or model artefacts.

References

Aicher S, Münzer A, Hezel J, Weber S (2023) Head pull-through capacity of load-bearing timber screws – Influential parameters and shortcomings of European test procedure. *Wood Material Science and Engineering.* DOI: 10.1080/17480272.2022.2155994.

ASTM F1575/F1575M (2021) *Standard test method for determining bending yield moment of nails.* ASTM International, West Conshohocken, PA, USA.

Aurand S, Blaß HJ (2021) Connections with inclined screws and increased shear plane friction. Paper 54-7-5. *INTER Meeting 54,* Online.

Azinović B, Frese M (2020) FE modelling of timber connections with inclined and cross-wise arranged screws – new findings on testing the shear stiffness. Paper 53-7-2. *INTER Meeting 53,* Online.

Bader TK, Schweigler M, Serrano E, Dorn M, Enquist B, Hochreiner G (2016) Integrative experimental haracterization and engineering modeling of single-dowel connections in LVL. *Construction and Building Materials* **107**:235-246. DOI: 10.1016/j.conbuildmat.2016.01.009.

Baesens B (2014) *Analytics in a big data world: The essential guide to data science and its applications.* ISBN 978-1-118-89271-8. Wiley & SAS business series, Hoboken, New Jersey.

Bejtka I (2003) Querzug- und Querdruckverstärkungen. *Karlsruher Tage – Forschung für die Praxis,* Karlsruhe, Germany.

Bejtka I, Blaß HJ (2002) Joints with inclined screws. Paper 35-7-4. *CIB-W18 Meeting 35,* Kyoto, Japan.

Bejtka I, Blaß HJ (2005) Self-tapping screws as reinforcements in connections with dowel-type fasteners. Paper 38-7-4. *CIB-W18 Meeting 38,* Karlsruhe, Germany.

Bejtka I, Blaß HJ (2006) Self-tapping screws as reinforcements in beam supports. Paper 39-7-2. *CIB-W18 Meeting 39,* Florence, Italy.

Blaß HJ (1990) Load distribution in nailed joints. Paper 23-7-2. *CIB-W18 Meeting 23,* Lisbon, Portugal.

Blaß HJ (1991) Traglastberechnung von Nagelverbindungen. *Holz als Roh- und Werkstoff* **49**(3):91-98. DOI: 10.1007/BF02614345.

Blaß HJ, Bejtka I, Uibel T (2006) *Tragfähigkeit von Verbindungen mit selbstbohrenden Holzschrauben mit Vollgewinde.* Karlsruher Berichte zum Ingenieurholzbau Band 4. Universität Karlsruhe (TH), Germany.

Blaß HJ, Bienhaus A, Krämer V (2000) Effective bending capacity of dowel-type fasteners. Paper 33-7-5. *CIB-W18 Meeting 33*, Delft, The Netherlands.

Blaß HJ, Colling F (2015) Load-carrying capacity of dowelled connections. Paper 48-7-3. *INTER Meeting 48*, Sibenik, Croatia.

Blaß HJ, Gard W (1994) Machine strength grading of timber. *Pacific Timber Engineering Conference*, Gold Coast, Australia.

Blaß HJ, Sandhaas C (2017) *Timber engineering. Principles for design.* ISBN 978-3-7315-0673-7. KIT Scientific Publishing, Karlsruhe.

Blaß HJ, Sandhaas C, Meyer N (2017) Steel-to-timber connections: Failure of laterally loaded dowel-type fasteners. Paper 50-7-1. *INTER Meeting 50*, Kyoto, Japan.

Blaß HJ, Steige Y (2018) *Steifigkeit axial beanspruchter Vollgewindeschrauben.* Karlsruher Berichte zum Ingenieurholzbau Band 34. Karlsruher Institut für Technologie.

Bocquet JF (1997) *Modélisation des deformations locales du bois dans les assemblages brochés et boulonnés.* Dissertation, University Blaise Pascal Clermont-Ferrand.

Bocquet JF, Barthram C, Pineur A (2012) L block failure of dowelled connections subject to bending reinforced with threaded rods. Paper 45-7-3. *INTER Meeting 45*, Växjö, Sweden.

Brandner R (2019) Properties of axially loaded self-tapping screws with focus on application in hardwood. *Wood Material Science and Engineering* **14**(5):254-268. DOI: 10.1080/17480272.2019.1635204.

Brandner R, Ringhofer A, Grabner M (2018) Probabilistic models for the withdrawal behaviour of single self-tapping screws in the narrow face of cross-laminated timber (CLT). *European Journal of Wood and Wood Products* **76**(1):13-30. DOI: 10.1007/s00107-017-1226-3.

Brandner R, Ringhofer A, Reichinger T (2019) Performance of axially loaded self-tapping screws in hardwood: Properties and design. *Engineering Structures* **188**:677-699. DOI: 10.1016/j.engstruct.2019.03.018.

Bratulic K, Augustin M, Schickhofer G (2019) Investigations concerning screw-press gluing of assemblies with CLT. Paper 52-18-1. *INTER Meeting 52*, Tacoma, USA.

CEN/TC 250/SC 5/WG 5 N 169 (2021) *Kessel Design and execution of stapled wood-based panel-to-timber connections for timber framed construction.* Comité Européen de Normalisation (CEN), Brussels, Belgium.

Chatterjee S, Price B (1995) *Praxis der Regressionsanalyse.* ISBN 3486229206. Oldenbourg, München Wien.

Chui YH, Craft S (2002) Fastener head pull-through resistance of plywood and oriented strand board. *Canadian Journal of Civil Engineering* **29**(3):384-388. DOI: 10.1139/l02-019.

Claus T, Seim W, Küllmer J (2022) Force distribution in self-tapping screws: experimental investigations with fibre Bragg grating measurement screws. *European Journal of Wood and Wood Products* **80**:183-197. DOI: 10.1007/s00107-021-01740-z.

Colling F (1990) *Tragfähigkeit von Biegeträgern aus Brettschichtholz in Abhängigkeit von den festigkeitsrelevanten Einflußgrößen*. Dissertation, Universität Karlsruhe (TH).

de Proft K (2017) *Properties of staples*. STSM report, Cost Action FP1402.

Delp A, Scholz R, Walther F, Mehlich D, Bletz-Mühldorfer O, Bathon L (2020) *Weiterentwicklung der Schraubenpressklebung im Holzbau*. Schlussbericht zu IGF-Vorhaben Nr. 19971N. Internationaler Verein für Technische Holzfragen, Germany.

Dias, AMPG (2005) *Mechanical behaviour of timber-concrete joints*. Dissertation, Delft University of Technology.

Dietsch P (2012) *Einsatz und Berechnung von Schubverstärkungen für Brettschichtholzbauteile*. Dissertation, Technische Universität München.

Dietsch P, Brandner R (2015) Self-tapping screws and threaded rods as reinforcement for structural timber elements – A state-of-the-art report. *Construction and Building Materials* **97**:78-89. DOI: 10.1016/j.conbuildmat.2015.04.028.

Dietsch P, Ringhofer A (2021) *Self-tapping screws as reinforcement for structural timber elements*. In: Branco J, Dietsch P, Tannert T (eds) RILEM State-of-the Art Reports 33, Reinforcement of timber elements in existing structures, pp. 7-27.

Dietsch P, Rodemeier S, Blaß HJ (2019) Transmission of perpendicular to grain forces using self-tapping screws. Paper 52-7-10. *INTER Meeting 52*, Tacoma, USA.

DIN 1052 (1969) *Holzbauwerke; Blatt 1: Berechnung und Ausführung*. Deutsches Institut für Normung (DIN), Berlin, Germany.

DIN 1052 (1988) *Holzbauwerke; Teil 2: Mechanische Verbindungen*. Deutsches Institut für Normung (DIN), Berlin, Germany.

DIN EN 1995 1-1 (2010) *Eurocode 5. Bemessung und Konstruktion von Holzbauten – Allgemeines – Allgemeine Regeln und Regeln für den Hochbau*. Comité Européen de Normalisation (CEN), Brussels, Belgium.

Dorn M, Habrová K, Koubek R, Serrano E (2021) Determination of coefficients of friction for laminated veneer lumber on steel under high pressure loads. *Friction* **9**:367-379. DOI: 10.1007/s40544-020-0377-0.

EAD 130118-00-0603 (2016) *Screws for use in timber constructions*. European Assessment document, EOTA, Brussels, Belgium.

EAD 130118-01-0603 (2019) *Screws and threaded rods for use in timber constructions.* European Assessment document, EOTA, Brussels, Belgium.

Eckert A, Brandner R, Glasner D (2023) Experimental study on control parameters for automated application and in-situ performance assessment of joints with self-tapping timber screws. *17th World Conference of Timber Engineering WCTE*, Oslo, Norway.

Eggersmann R, Kirchdoerfer T, Reese S, Stainier L, Ortiz M (2019) Model-free data-driven elasticity. *Computer Methods in Applied Mechanics and Engineering* **350**:81-99. DOI: 10.1016/j.cma.2019.02.016.

Ehlbeck J (1979) *Nailed joints in wood structures.* Bulletin no. 166. Wood Research and Wood Construction Laboratory, Virginia Polytechnic Institute and State University, USA.

Ehlbeck J, Siebert W (1984) *Tragverhalten von Nagelverbindungen bei gleichzeitiger Beanspruchung auf Abscheren und Ausziehen.* Forschungsbericht der Versuchsanstalt für Stahl, Holz und Steine. Universität Karlsruhe (TH), Germany.

EN 1382 (2016) *Timber structures – Test methods – Withdrawal capacity of timber fasteners.* Comité Européen de Normalisation (CEN), Brussels, Belgium.

EN 1383 (2016) *Timber structures – Test methods – Pull through resistance of timber fasteners.* Comité Européen de Normalisation (CEN), Brussels, Belgium.

EN 14080 (2013) *Timber structures – Glued laminated timber and glued solid timber. Requirements.* Comité Européen de Normalisation (CEN), Brussels, Belgium.

EN 14358 (2016) *Timber structures – Calculation and verification of characteristic values.* Comité Européen de Normalisation (CEN), Brussels, Belgium.

EN 14592 (2012) *Timber structures – Dowel-type fasteners – Requirements.* Comité Européen de Normalisation (CEN), Brussels, Belgium.

EN 15737 (2009) *Timber structures – Test methods – Torsional resistance of driving in screws.* Comité Européen de Normalisation (CEN), Brussels, Belgium.

EN 1993 1-1 (2010) *Eurocode 3. Design of steel structures – Part 1-1: General rules and rules for buildings.* Comité Européen de Normalisation (CEN), Brussels, Belgium.

EN 1993 1-8 (2010) *Eurocode 3. Design of steel structures – Part 1-8: Design of joints.* Comité Européen de Normalisation (CEN), Brussels, Belgium.

EN 1995 1-1 (2010) *Eurocode 5. Design of timber structures – Part 1-1: General – Common rules and rules for buildings.* Comité Européen de Normalisation (CEN), Brussels, Belgium.

EN 1995 1-1/A2 (2014) *Eurocode 5. Design of timber structures – Part 1-1: General – Common rules and rules for buildings.* Comité Européen de Normalisation (CEN), Brussels, Belgium.

EN 26891 (1991) *Timber structures – Joints made with mechanical fasteners – General principles for the determination of strength and deformation characteristics (ISO 6891)* Comité Européen de Normalisation (CEN), Brussels, Belgium.

EN 338 (2016) *Structural timber – Strength classes.* Comité Européen de Normalisation (CEN), Brussels, Belgium.

EN 383 (2007) *Timber structures – Test methods – Determination of embedment strength and foundation values for dowel-type fasteners.* Comité Européen de Normalisation (CEN), Brussels, Belgium.

EN 409 (2009) *Timber structures – Test methods – Determination of the yield moment of dowel type fasteners.* Comité Européen de Normalisation (CEN), Brussels, Belgium; Belgium.

EN ISO 8970 (2010) *Timber structures – Testing of joints made with mechanical fasteners – Requirements for wood density.* Comité Européen de Normalisation (CEN), Brussels, Belgium.

ENV 1995 1-1 (1993) *Eurocode 5. Design of timber structures – Part 1-1: General – Common rules and rules for buildings.* Comité Européen de Normalisation (CEN), Brussels, Belgium.

Filiatrault A, Folz B (2002) Performance-based seismic design of wood framed buildings. *ASCE Journal of Structural Engineering* **128**(1):39-47. DOI: 10.1061/(ASCE)0733-9445 (2002)129:1(39).

Foschi RO (1974) Load-slip characteristics of nails. *Wood Science* **7**(1):69-76.

Franke S (2008) *Zur Beschreibung des Tragverhaltens von Holz unter Verwendung eines photogrammetrischen Messsystems.* Dissertation, Bauhaus-Universität Weimar.

Franke S, Magnière N (2014) Discussion of testing and evaluation methods for the embedment behaviour of connections. Paper 47-7-1. *INTER Meeting 47*, Bath, UK.

Frese M (2019) Density variations in beech LVL: Influence on insertion moment and withdrawal capacity of screws. Paper 52-7-3. *INTER Meeting 52*, Tacoma, USA.

Frese M, Blaß HJ (2017) *Entwicklung von selbstbohrenden Schrauben für Laubholz höherer Dichte (FAGUS)* Schlussbericht für die Öffentlichkeit FKZ 031A437J, Karlsruher Institut für Technologie.

Frese M, Fellmoser P, Blaß HJ (2010) Modelle für die Berechnung derAusziehtragfähigkeit von selbstbohrenden Holzschrauben. *European Journal of Wood and Wood Products* **68**(4):373-384. DOI: 10.1007/s00107-009-0378-1.

Fürst H (2019) *Überprüfung und Optimierung der Regelungen für streifen- und plattenförmige Schraubpressverklebungen in ÖNORM B-1995-1-1.* Master thesis, Technische Universität Graz.

Hartung J (2009) *Statistik. Lehr- und Handbuch der angewandten Statistik.* ISBN 978-3-486-59028-9. Oldenbourg, München, 15. Auflage.

Hölz KS (2021) *Analyse und Modellierung der Zusammenhänge zwischen der Tragfähigkeit und den Gewindeparametern von Holzschrauben durch die Entwicklung von Analysetechniken.* Dissertation, Karlsruher Institut für Technologie.

Hübner U (2013) *Mechanische Kenngrößen von Buchen-, Eschen- und Robinienholz für lastabtragende Bauteile.* Dissertation, Technische Universität Graz.

ISO 8970 (1989) *Timber structures – Testing of joints made with mechanical fasteners – Requirements for wood density.* International Organization for Standardization, Geneva, Switzerland.

Jockwer R (2016) Impact of varying material properties and geometrical parameters on the reliability of shear connections with dowel-type fasteners. Paper 49-7-1. *INTER Meeting 49*, Graz, Austria.

Jockwer R, Dietsch P (2018) Review of design approaches and test results on brittle failure modes of connections loaded at an angle to the grain. *Engineering Structures* **171**:362-372. DOI: 10.1016/j.engstruct.2018.05.061.

Johansen KW (1949) *Theory of timber connections.* Publication 9. International Association of Bridge and Structural Engineering (IABSE), Basel, Switzerland.

Jorissen A (1998) *Double-shear timber connections with dowel-type fasteners.* Dissertation, Delft University of Technology.

Khan R, Niederwestberg J, Chui YH (2021) Influence of insertion angle, diameter and thread on embedment properties of self-tapping screws. *European Journal of Wood and Wood Products* **79**:707-718. DOI: 10.1007/s00107-020-01651-5.

Knorz M, Van de Kuilen JWG (2012) Development of a high-capacity engineered wood product – LVL made of European beech (Fagus sylvatica L.) *12th World Conference of Timber Engineering WCTE*, Auckland, New Zealand.

Kuck E, Sandhaas C (2022) Load-bearing behaviour of partially threaded screws in hardwood. *Wood Material Science and Engineering.* DOI: 10.1080/17480272.2022.2084453.

Kuenzi EW (1955) *Theoretical design of a nailed or bolted joint under lateral load.* Report D1951. Forest Products Laboratory, Madison, Wisconsin, USA.

Kumpenza C, Ringhofer A, Krenke T, Sotayo A, Pramreiter M, Müller U (2020) Timber screw connection: Study of the strain along the interface using optical measurement techniques and simulations. *Bioresources* **15**(2):3859-3873. DOI: 10.15376/biores.15.2.3859-3873.

Leijten AJM (2009) Withdrawal capacity of washers in bolted timber connections. *Wood Material Science and Engineering* **4**(3-4):131-139. DOI: 10.1080/17480270903411066.

Leijten AJM, Jorissen A, Leijer B de (2012) The local bearing capacity perpendicular to grain of structural timber elements. *Construction and Building Materials* **27**(1):54-59. DOI: 10.1016/j.conbuildmat.2011.07.022.

Lemaître R (2020) *Développement d'un outil de calcul non linéaire de dimensionnement d'assemblages bois tridimensionnels soumis à des torseurs plans*. Dissertation, Université de Lorraine.

Lemaître R, Schweigler M, Bader TK (2023) Combined embedment and axial load tests of a steel dowel in various wood species. *17th World Conference of Timber Engineering WCTE*, Oslo, Norway.

Liu WK, Karniadakis G, Tang S, Yvonnet J (2019) A computational mechanics special issue on: data-driven modeling and simulation—theory, methods, and applications. *Computational Mechanics* **64**(2):275-277. DOI: 10.1007/s00466-019-01747-7.

Mahlknecht U, Brandner R, Ringhofer A (2021) Minimum geometric and execution requirements for axially loaded groups of screws in hardwood. Paper 54-7-6. *INTER Meeting 54*, Online.

Marsden JR, Pingry DE (2018) Numerical data quality in IS research and the implications for replication. *Decision Support Systems* **115**:A1-A7. DOI: 10.1016/j.dss.2018.10.007.

Mestek P (2011) *Punktgestützte Flächentragwerke aus Brettsperrholz (BSP) – Schubbemessung unter Berücksichtigung von Schubverstärkungen*. Dissertation, Technische Universität München.

Moeller T (1951) *En ny metod foer beraekning av spikfoerband.* Handlingar no. 117. Chalmers Tekniska Hoegskola, Göteborg, Sweden.

Möhler K, Ehlbeck J, Köster P (1973) *Untersuchungen über das Trag- und Verformungsverhalten von Heftklammerverbindungen bei Tafelelementen.* Mitteilung aus der Versuchsanstalt für Stahl, Holz und Steine. Universität Karlsruhe (TH).

Munch-Andersen J, Sorensen JD (2011) Pull-through capacity in plywood and OSB. Paper 44-7-1. *INTER Meeting 44*, Alghero, Italy.

Munch-Andersen J, Svensson S (2016) Accurate strength parameters for fasteners with examples for ring shank nails. *14th World Conference of Timber Engineering WCTE*, Vienna, Austria.

NZS AS 1720.1 (2018) *Timber Structures. Part 1: Design methods.* Draft Number DZ NZS AS 1720.1/V6.0, Standards New Zealand.

Ogawa K, Harada M, Shibusawa T, Miyamoto K (2019) Method for measuring the resistances produced on parallel and perpendicular veneers in plywood under nail embedment loading. *Journal of Wood Science* **65**:1. DOI: 10.1186/s10086-019-1786-4.

Ottenhaus L-M, Jockwer R, van Drimmelen D, Crews K (2021) Designing timber connections for ductility – A review and discussion. *Construction and Building Materials* **304**:124621. DOI: 10.1016/j.conbuildmat.2021.124621.

Ottenhaus L-M, Li Z, Crews K (2022) Half hole and full hole dowel embedment strength: A review of international developments and recommendations for Australian softwoods. *Construction and Building Materials* **344**:128130. DOI: 10.1016/j.conbuildmat.2022.128130.

Pirnbacher G, Brandner R, Schickhofer G (2009) Base parameters of self-tapping screws. Paper 42-7-1. *CIB-W18 Meeting 42*, Dübendorf, Switzerland.

prEN 14592 (2017) *Timber structures – Dowel-type fasteners – Requirements.* Comité Européen de Normalisation (CEN), Brussels, Belgium.

Radhakrishna Rao C, Toutenburg H, Shalabh, Heumann C (2008) *Linear models and generalisation. Least squares and alternatives. Chapter 8: Analysis of incomplete data sets.* Springer Series in Statistics, Springer, Heidelberg. DOI 10.1007/978-3-540-74227-2_8.

Ringhofer A (2017) *Axially loaded self-tapping screws in solid timber and laminated timber products*. Dissertation, Graz University of Technology.

Ringhofer A, Schickhofer G (2014) *Investigations concerning the force distribution along axially loaded self-tapping screws.* In: Aicher S, Reinhardt H W, Garrecht H (eds) RILEM bookseries. Materials and joints in timber structures. Recent developments in technology. Stuttgart, Germany, pp. 201-210.

Rodd PD (1973) *The analysis of timber joints made with circular dowel connectors*. Dissertation, University of Sussex.

Sandhaas C (2012) *Mechanical behaviour of timber joints with slotted-in steel plates*. Dissertation, Delft University of Technology.

Sandhaas C, Blaß HJ (2021) Steel properties of self-tapping screws. Paper 54-7-1. *INTER Meeting 54*, Online.

Sandhaas C, Blaß HJ (2022) Head pull-through properties of self-tapping screws. Paper 55-7-1. *INTER Meeting 55*, Bad Aibling, Germany.

Sandhaas C, Görlacher R (2018) Analysis of nail properties for joint design. *Engineering Structures* **173**:231-240. DOI: 10.1016/j.engstruct.2018.06.071.

Sandhaas C, Ravenshorst G, Blaß HJ, Van de Kuilen JWG (2013) Embedment tests parallel-to-grain and ductility aspects using various wood species. *European Journal of Wood and Wood Products* **71**(5):599-608. DOI: 10.1007/s00107-013-0718-z.

Sartori T, Tomasi R (2013) Experimental investigation on sheathing-to-framing connections in wood shear walls. *Engineering Structures* **56**:2197-2205. DOI: 10.1016/j.engstruct.2013.08.039.

Schweigler M, Bader TK, Bocquet JF, Lemaître R, Sandhaas C (2019) Embedment test analysis and data in the context of phenomenological modeling for dowelled timber joint design. Paper 52-7-8. *INTER Meeting 52*, Tacoma, USA.

Schweigler M, Bader TK, Vessby J, Eberhardsteiner J (2017) Constrained displacement boundary condition in embedment testing of dowel-type fasteners in LVL. *Strain* **53**:e12238. DOI: 10.1111/str.12238.

Schweigler M, Vedovelli M, Lemaître R, Bocquet JF, Sandhaas C, Bader TK (2021) Beam-on-foundation modeling as an alternative design method for timber joints with dowel-type fasteners - part 3: Second order theory effects for considering the rope effect. Paper 54-7-8. *INTER Meeting 54*, Online.

Sjödin J, Serrano E, Enquist B (2008) An experimental and numerical study of the effect of friction in single dowel joints. *Holz als Roh- und Werkstoff* **66**(5):363-372. DOI: 10.1007/s00107-008-0267-z.

Stamatopoulos H, Malo KA (2015) Withdrawal capacity of threaded rods embedded in timber elements. *Construction and Building Materials* **94**:387-397. DOI: 10.1016/j.conbuildmat.2015.07.067.

Steiger R, Köhler J (2005) Analysis of censored data - examples in timber engineering research. Paper 38-17-1. *CIB-W18 Meeting 38*, Karlsruhe, Germany.

Steilner M (2014) Pre-stressing of wood with full thread screws. *COST Action FP1004 Conference on Experimental Research with Timber*, Prague, Czech Republic.

Steilner M, Blaß HJ (2014) *A method to determine the plastic bending angle of dowel-type fasteners*. In: Aicher S, Reinhardt H W, Garrecht H (eds) RILEM bookseries. Materials and joints in timber structures. Recent developments in technology. Stuttgart, Germany, pp. 603-613.

Steilner M, Kunkel H, Sandhaas C (2022) Low cycle ductility of self-tapping screws. Paper 55-7-5. *INTER Meeting 55*, Bad Aibling, Germany.

Svensson S, Munch-Andersen J (2018) Theory of timber connections with slender dowel type fasteners. *Wood Material Science and Engineering* **13**(1):7-15. DOI: 10.1080/17480272.2016.1226382.

Taylor SE, Bender DA (1989) A method for simulating multiple correlated lumber properties. *Forest Products Journal* **39**(7/8):71-74.

Uhl M, Bruchmüller T, Matthiesen S (2019) Experimental analysis of user forces by test bench and manual hammer drill experiments with regard to vibrations and productivity. *International Journal of Industrial Ergonomics* **72**:389-407. DOI: 10.1016/j.ergon.2019.06.016.

Vedovelli M (2020) Untersuchungen zu Zugscherverbindungen in Laubholz. *Doktorandenkolloquium*, Stuttgart, Germany.

Werner H (1993) *Untersuchungen von Holz-Verbindungen mit stiftförmigen Verbindungsmitteln unter Berücksichtigung streuender Einflußgrößen*. Dissertation, Universität Karlsruhe (TH).

Werner H, Siebert W (1991) Neue Untersuchungen mit Nägeln für den Holzbau. *Holz als Roh- und Werkstoff* **49**(5):191-198.

Westermayr M, van de Kuilen JW (2020) A conceptual model to predict the withdrawal capacity of screws inserted parallel to grain in beech, ash and spruce. Paper 53-7-1. *INTER Meeting 53*, Online.

Whale LRJ, Smith I (1986a) *Mechanical timber joints.* Report 18/86. TRADA, High Wycombe, UK.

Whale LRJ, Smith I (1986b) The derivation of design clauses for nailed and bolted joints in Eurocode 5. Paper 19-7-6. *CIB-W18 Meeting 19*, Florence, Italy.

White MS, Gales TL (1990) *Quality variations in helically threaded nails.* Pallet and Container Laboratory Bulletin No. 15. Virginia Polytechnic Institute and State University, U.S.A.

Wilkinson TL (1971) Theoretical lateral resistance of nailed joints. *Journal of the Structural Division* **97**:1381-1398.

Windeck L, Blaß HJ (2017) Querdruckverhalten von Brettschichtholz aus Nadelholz. *Bautechnik* **94**(11):767-775. DOI: 10.1002/bate.201700048.

Yurrita M, Cabrero JM (2019) New analytical model for brittle failure in the parallel-to-grain direction of timber connections with large diameter fasteners. Paper 52-7-7. *INTER Meeting 52*, Tacoma, USA.

Yurrita M, Cabrero JM (2021) New analytical model for plug shear of timber connections with small diameter dowel-type fasteners in the parallel-to-grain direction. Paper 54-7-10. *INTER Meeting 54*, Online.

Zarnani P (2013) *Load-carrying capacity and failure mode analysis of timber rivet connections*. Dissertation, The University of Auckland.